On the Smell of

LANDSMAN'S BOOK SHOP LTD.
Bromyard, Herefordshire

On the Smell of an Oily Rag

my 50 years in farming

John Cherrington

FARMING PRESS

First published 1979
First paperback edition 1993

Copyright © 1979 and 1993
the Executors of John Cherrington's Estate

All rights reserved. No parts of this
publication may be reproduced, stored in
a retrieval system, or transmitted, in any
form or by any means, electronic, mechanical,
photocopying, recording or otherwise, without
prior permission of Farming Press Limited

The publisher gives thanks for permission to
reproduce the photographs from the Institute of
Agricultural History and Museum of English
Rural Life (MERL), University of Reading, or as
otherwise acknowledged

ISBN 0 85236 256 0

A catalogue record for this book is available
from the British Library

Published by Farming Press Books
Wharfedale Road, Ipswich IP1 4LG, United Kingdom

Distributed in North America
by Diamond Farm Enterprises,
Box 537, Alexandria Bay, NY 13607, USA

Cover design by Andrew Thistlethwaite
Typesetting by Galleon Photosetting, Ipswich
Printed and bound in Great Britain by
Biddles Ltd, Guildford and King's Lynn

Contents

	Preface	vii
1	Early Days	1
2	Decision Time	6
3	Emigration	19
4	Glencairn	28
5	Shearing and Threshing	40
6	Beef in Argentina	48
7	Back in England	68
8	Manor Farm, Knoyle	76
9	Greens Farm, Chute	93
10	The Docks at Tangley	108
11	The Outbreak of War	127
12	Stock Development	145
13	Sheep Farming	151
14	Wartime Arable	159
15	I Become a Landowner	168
16	A Question of Finance	178
17	Outside Interests	185
18	Farming Politics	194
19	I Break into Journalism	203

20	Where is Farming Going?	214
21	Lessons from Abroad	223
22	Feeding the World	232
23	Family and Friends	241
24	The Essential Factor	247

*A section of photographs appears between
pages 152 and 153*

Preface

There is always a period before the recent past becomes history, but when the story of British agriculture is finally written there is no doubt that one of its golden ages will be the period from the 1930s to the 1980s.

Written in 1979, John Cherrington's autobiography spans this 50 years – from the Great Depression – through the war years when farming took on the challenge of feeding the nation – to post-war prosperity and then the Common Market.

John Cherrington, who died in 1988, was well known as a journalist, broadcaster, author and practical farmer, and he was never afraid to write exactly what he thought. There was no malice in his sharp observations and more often than not he was the butt of his own humour.

This autobiography gives a unique picture of farming in a time of great change when it was possible to work your way to the very top of the industry despite starting off with little more than the smell of an oily rag.

St Weonards, Hereford DAN CHERRINGTON
March 1993

1. Early Days

My formative years, those that had the most influence on my approach to farming, and much of my attitude to life were spent in New Zealand and Argentina. Times in New Zealand were hard. Prices far lower than any I had seen in England in the late 1920s, and have seen since. The farmers, among whom I worked, were mostly in debt, some were being sold up. Yet in every community there was the odd individual who seemed to be getting on with luck – you always need a lot of luck – and a blend of parsimony and farming competence. It was not that they were particularly good farmers. Their talent seemed to be to achieve results with the least possible effort and the expenditure of the minimum of cash. It was not tidy farming, it was not as productive as that of many of their neighbours, but it meant solvency.

Such people were looked down on by the more traditional, and were generally described as being able to thrive 'on the smell of an oily rag'. The first New Zealander I worked for was certainly not one of these. Not long married, he had, even in my immature judgement, his priorities wrong. For three days one summer we laboured. Not in fencing, as we should have, the paddock where his sale ewes were to be held before the drover came to take them away; but in casting a flight of steps in concrete from the verandah to the garden.

The inevitable happened. We gathered the sheep, drafted out the sale ewes, and left them overnight in the paddock from which most of them proceeded to escape back to the most distant corner of the run. By the time we had remustered them and got them safely penned, prices had collapsed. This was a blow from which he never recovered and eventually he had to give up the farm.

I saw those steps thirty years later, the only permanent feature of the decaying wooden farmhouse now uninhabited; our names and the date were still readable. If we had not built them the ewes would have made another £500 and, who knows, he might have survived. I pointed them out to my wife so that she might understand my life-long aversion to gardening.

This experience, coupled with an almost paranoiac fear of unemployment and insolvency, has been a major factor in my farming. It is, I would be the first to admit, a most negative approach. But I was brought up in the 1920s and early 1930s, in an atmosphere of unemployment, slump and depression. There were up to 3,000,000 unemployed in Britain, and 30,000 in New Zealand in a population of about 1,000,000 at the time. Living in the Southern English counties and in New Zealand I did not see the worst of the industrial misery. Even so the atmosphere bred a determination to avoid getting in the same boat.

Of course I may have been mean by nature, and the environment of those days simply fostered a natural characteristic, which in any case was essential for survival in the first dozen or so years of my farming: indeed until the war came to alter the whole concept of farming and everything else for at least a generation. But by then the habits had been acquired and there was no changing them. I am by any count prosperous now, but I still begrudge any farming activity which does not directly lead to profit.

I am grateful to New Zealand for having broken me of the dreadful class consciousness with which home and school had infected me. My parents were not rich; my father was a banker who had worked himself up from clerk to general manager by hard work and the ability to take decisions. My mother was a considerable and overwhelming influence on my life. Both had had childhoods where money was short and were climbing out of the lower to the upper middle classes.

They were both Liberals, and I remember now the sadness with which they confessed by letter that, in the 1931 election, they had deserted the Liberals to vote Conservative. Not because they liked them, but because they felt it was for the country's good. I think in the end they regretted it. But Liberal though they might have been, the stratifications of class were paramount. The friends I made at school would not be asked to tea unless their mothers had called or been called on.

Newcomers to the town waited for callers who would drop their cards on a silver tray and, if the contact turned out satisfactory, would extend their friendship. A newcomer must never, I understood, call first on an older resident. Some of my school friends had parents who kept shops in the town. These were 'in trade' and were never invited. As far as I can gather, they never expected to be. I remember going to see them guiltily at the back of the shop.

Nor was friendship encouraged with the self-made – people in trade who were better off than my parents, and who had cars long before we did. We might meet them daily at school but, although we lived only a few streets away, holiday attachments were, until I was already old enough to leave school, negligible and only carried out under a barrage of silent disapproval. It was not, my mother told me when I questioned her on these things, that these people were in any way unworthy. It was just that they had nothing in common with us and that to achieve too great an intimacy would be embarrassing to both sides.

The fear behind it all was that I would fall in love with an 'unsuitable' (in the social sense) young woman, marry her and lead a life of perpetual embarrassment, breeding more embarrassments. For this reason the social life of adolescents still at home was conducted more carefully then than I manage my pig herd today. Children of selected families met at the same dances and, as far as was humanly possible, were kept to these circles.

From about the age of twelve I boarded at the local school, Berkhamsted. It was already famous for the fact that Graham Greene, the headmaster's son, had been educated there and, as far as we could gather, rebelled against it. I hadn't the guts to rebel, and sullenly put up with the most miserable four years I have spent in my life. The school accepted a proportion of scholarship boys and many of these who came by train were segregated in a separate house with the unfortunate name of Addlebert or Adders. This may have been a matter of convenience, but the unwritten apartheid made absolutely certain that there would be the minimum intercourse between those whose parents paid the fees, and those who had scholarships.

I have never understood the purpose of the incarceration of children from an early age in the institutional surroundings of the average public school of those days. I had been a day boy elsewhere previously; perhaps if I had been through prep school since the age of seven or eight, as had many of my fellows, I might have been conditioned into putting up with boarding better. As it was the loss of freedom to roam the streets, the lack of privacy, and the conformity in behaviour which was rigorously enforced by the mass harassment of my school fellows, made me angry and unhappy until, I am ashamed to say, I joined them in prejudice and custom.

I am still ashamed of my behaviour in those years, and wish that I had had the strength of character to stand out against them, or just to walk out of the school. I now realise that would have been the simplest way out. In the end I came to conform, to the extent of co-operating with some of the masters. Being fairly good at rugger, I could revenge myself on my enemies in the scrum, under the pretext of sport.

I had one great disappointment, which probably altered the course of my life. I was getting quite interested in science and biology but, when I was fourteen or so, there was a selection process with the choice of specialising in Science, the Arts, i.e. English and languages, or Classics (Latin and Greek). I chose science but could not get my mother's approval. She was obsessed by the notion that scientists were lower-class narrow-minded people, not much above mechanics. To prove her point she instanced that none of my science masters spoke with what used to be called a public school accent. So once more I did as she told me.

I am sure now that, had I taken science (I have a good practical mind), I could have become a doctor; and doctors I believe are among the few who in their daily lives and work do their very best for the rest of mankind. There are, I know, unskilled doctors, but I know few who are deliberately slack and careless.

The only advantage I really gained from school, besides a friendship or two, was the ability to put up with bad food, discomfort and harassment. Nothing I have experienced since, here or abroad, has ever been as bad, or could be, as my years at Berkhamsted.

2. Decision Time

I am not sure when I decided to be a farmer. I knew nothing of the trade, and had only stayed on farms on holiday when I had taken little part in what was going on. The sort of farms where lodgers were taken were usually those whose owners were on the downward path. This is not so today when holidaymakers are a business and almost certainly a source of tax-free pocket money. I think the notion came to me that a farmer was his own boss, and could do what he liked. I disliked and resented authority in any form, and I guessed that employers were just like schoolmasters or parents but could sack you into the bargain. I wanted nothing to do with them.

My parents took my decision with good grace. My father realised I was not made for the city, and my mother accepted the fact that I would never be an academic. But they insisted that I should go to Cambridge to get an agricultural degree. They set great store by this; none of the family had been to Oxbridge and, as the eldest son whose father could afford the fees, they were certain I should go. I was just as convinced that I should not.

At the time and for many years, I was sure that I had made the right decision. Cambridge I thought would be simply an expensive extension of school. During the endless arguments that persisted for quite a long time, I repeated the well-voiced criticism by the few farmers I had met that the only place to learn farming was on the farm and not in college. This attitude persists today and, in most respects, it is right.

I also made the point that, as the family capital was

short, any money that would be saved by avoiding university fees should be my starting capital. It is quite possible that had I succumbed to their arguments I would have left the Agricultural Department with a minor degree, no capital and a fear of risking the plunge into practical farming.

Had this been the case, I would probably have taken up a safe profession with a pension at the end of it, and by now would be retired and gradually fading from the scene. I would have been too old at 21 or so to face starting farming from the bottom. On the other hand, there is no doubt that some who go to a university do manage to widen their horizons in a manner which gives them considerable advantages.

Two of the best farmers I know are both graduates; but one came from a farming background and the other is undoubtedly wasting his very considerable talents on a 500 acre farm. However both had ample capital and neither studied agriculture in any shape or form until they actually became involved in it, when well into their thirties.

After winning the argument over Cambridge my parents insisted that I should be made to learn a skill which could assist me in the farming line. This was not to be ploughing, milking or other elemental essential task, but woolsorting. The reason for this was my determination to go abroad. Australia was the choice, but because relatives both there and in New Zealand said what a rough place the former was, New Zealand it was going to have to be; and New Zealand was full of sheep. I concurred in this, all the newspapers were full of the opportunities to be found in these new countries; and I had recently watched the All Blacks demolish all opposition.

British farming was, as ever, said to be in a sad decline, and many young men were contemplating emigration rather than try their luck at home. In this, as events were to prove, they were abundantly wrong. The only place in which to farm during a time of depression is close to the market. I found things were many times worse abroad.

I had no inkling of this at the time, but in any case wanted to get as far away from the constraints of home and family as I could. My parents considered that I was too young to emigrate, so I consented to a compromise of six months on a farm and then a sort of sandwich course at Leeds University studying textiles. It was hoped that once I had got involved in the wool trade I might remain there. Some of my father's customers were successful woolmen, and undoubtedly there could have been useful connections in that quarter.

My farming start was inauspicious. I was taken as prospective pupil to one of the best farms in Suffolk, and all one Saturday my father and I were shown round the sheep, the cattle, the sugar-beet and the cauliflowers and so on. The farmer appeared to be a rather overawing character, but I would probably have got on with him and most certainly would have learnt a lot.

During tea in the farmhouse, it was agreed that I should go there, and we drove home with Father obviously relieved to have got me settled so easily. His relief was short-lived. Mother cross-examined us on our arrival home, and all went well until we mentioned that there were three daughters, to whom I had been introduced. 'In that case', she said, 'he is not to go there'. I protested that I was not interested in girls, and even if I were, they were not my particular type. Frankly I thought that in their school uniforms they were rather plain. 'Beauty,' said my mother, 'has nothing to do with it. I know what propinquity can do.'

It was many years before I found what propinquity meant: that on a desert island or in an isolated farmhouse, the most unlikely young couple will fall for each other whatever their looks or characters. My mother probably saved me from becoming an orthodox East Anglian farmer, whereby I missed a great deal of experience.

I was telling this story at a farmers' lunch in Salisbury well after the war when a member of the group interrupted saying that he too had been taken to a Suffolk farm. We compared notes – it was the same and he had married one of the daughters. We met at a dance shortly afterwards and for old times' sake I took his wife around the floor. We were far from compatible, and in this case propinquity would have made a blunder.

My father, who was in his quiet way a pretty decisive character, very soon settled me. He knew a farmer in Shropshire, a client of his bank, who I suspect was not in a position at the time to argue with his bank manager.

I was sent to Mr P.G. Holder, as they say sight unseen, for a fee of £50 to learn farming for six months.

At the time of my engagement, Holder lived in great style in a large country house four miles from Craven Arms. I remember being driven there by my parents. With some foreboding I marched with them up the steps to knock on the front door. It was opened by what, in those days, would have been described as a flapper with high heels, short skirts (not quite a mini) and a very soignée get-up altogether for three o'clock of a country afternoon. Her first words were 'You must be the new pupil; do you do the flat Charleston?'

My mother was a strong-minded woman, but this was the only time I ever saw her jaw drop. The die was cast, however, and she could not very well bring propinquity into it at that stage. After a rather stiff tea-party, I was left to a fate which I was to enjoy enormously.

P.G. Holder was, in almost all respects, an extraordinary man. The third son of a medium-sized farmer near Tenbury in Worcestershire, he was articled to a local solicitor at the age of 15. He still practised for a few clients while I knew him, but his real passion was for farming and land speculation. In many respects he was not much of a farmer. He had a good eye for stock and an understanding of arable farming, but he had no idea of farming as a business. Where he excelled was in land speculation. From it he made several fortunes, all but one of which he subsequently lost in farming. Nonetheless he certainly enjoyed his life; he was completely governed by impulse and enthusiasm.

If a property was up for sale he would leaf through the particulars, noting what he thought he could sell the various houses, farms and woods for. Then, if the sums looked right, he bid at the auction. If the property was knocked down to him he wrote out the cheque for the deposit, and then – only then – told the bank that he might have to borrow some money if he failed to sell out before completion, as he often could not.

I once asked him if he consulted his banker before he made a bid. 'Oh dear no!' was his reply. 'No bankers understand business as it is conducted in the real world. Never tell your banker any more than you actually have to. Remember they only supply money, just as grocers do tea and butter. You don't tell grocers your business, do you?'

I have never been able to adopt his high-handed attitude to bank managers, whose only defence in his case was to refer the case to their boss, who was my father. Here Greek must have met Greek with a vengeance, with P.G. giving away as little as possible while my father would warn him against the dangers of overinvestment in speculative deals with other people's money. However it wasn't the speculation that lost money, but the farming. This was because P.G. was a man of ideas. Not for him a system which was long-term and depended on the slow passage of the seasons, or not if he could help it.

When I joined him he was in the cattle-fattening business. Much of his land was in the Corve Dale near Ludlow, the traditional fattening land for Hereford cattle.

So all that spring we went to markets in the Welsh Border counties: Hereford, Leominster, Ludlow, Knighton, Bishops Castle and many others. There he bought heifers, because he was a heifer grazier, probably because they finish rather faster than steers. There were no lorries in those days, and the purchases were either railed to Craven Arms and walked from there, or were walked back from the markets by drovers who were paid around 6d (2$\frac{1}{2}$p) an hour. Many times I helped to drive them back myself and spent days and nights tramping the summer lanes.

Once home these cattle were fed a concentrated feed as well as the grass, and within a few weeks were judged ready for market and they retraced their steps. I have always had a good memory for cattle and soon noticed that the animals I was taking to market were making less than they had cost some weeks before. I pointed this out to him and he told me that although individual cattle lost money, the bulk of the enterprise paid. 'But Mr Holder,' I replied 'how can the whole pay, when individually each animal loses money?'

'Well,' he said, 'you may be right, but it's my money, my cattle and my farm. I started with nothing and am far from bankrupt yet, and what is more I enjoy my life'.

This most obviously was not the farm for me to learn on in any orthodox sense and I should have been far better instructed in East Anglia, even if propinquity had done its worst. But although with Holder I never learnt to plough, sow, reap or mow. I did, in an entirely negative way, learn the elements of business. Even now, when faced with a farming problem, I ask myself would P.G. have done it this way? If he would, then it was not for me.

The farm was littered with his failed hopes – embryo schemes which had come unstuck. There were four miles of light railway, ex World War I surplus, which were to be laid from Craven Arms to the fruit farm, mainly plums, which he had established. Three 1912 cars were in the garage waiting for prices to rise. He converted ex-army lorries for livestock carrying a year or two too soon. He never seemed to have less than half a dozen huge country houses on his hands – the remnants of estates he had split up. All this collection was still on the farm when war broke out and was mostly sold for scrap at a considerable loss.

The plums were a perpetual nightmare, to me at any rate, as I was left partly in charge of them as a pupil and wholly so when I worked for him some years later when they were being harvested. Plums are only pickable and transportable for a very short time; they were also unpopular in those days. The British forces in World War I had been saturated with plum and apple jam, and they and their families did not want to know them. So picking them when just ripe was a problem and selling them was worse. P.G. himself would take a half-ton trailer behind his car and set off at midnight to tour the Welsh coast towns in the early morning to sell to small greengrocers.

I was left behind to sell to the representatives of the chain stores and the dealers. I remember the Marks and Spencer buyer giving me $\frac{1}{2}d$ (0.35p) a pound in their baskets for many tons. However it was better than letting them go bad. In later years, because I resolutely refused to leave my farm for the fruit season and help him out, he kept a herd of pigs in the orchard. These kept knocking the trees and the ripe fruit would fall for them. If there was a customer, the pigs were moved out and picking could begin.

However, I enjoyed the summer in Shropshire. The daughters were delightful company and being older for the most part were not interested in me except as a driver or dance partner when nothing else was available. Mrs Holder was apt to conscript me for tennis, a game I hated and avoided as much as possible. I also played some village cricket. Such work as I did was all in the fashion of the squire's son, counting cattle grazing in the fields; shooting pigeons, rooks and rabbits; a little haymaking or harvesting. Several pupils had preceded or followed me – none of them succeeded in farming that I can remember.

I then spent about six months at Leeds University during which time I took in the basis of a course in textiles, followed by some months in a woollen mill learning wool-sorting. The idea behind this was ostensibly to ground me for New Zealand sheep-farming, by giving me a knowledge of wool. It was also an endeavour on the part of my parents to persuade me to go into some nice safe occupation like the wool trade, which at that time was booming.

It was, of course, the very worst preparation for a life in practical farming. What a boy, or girl, for that matter needs is to get the feel of farming, an eye for stock or crops and the state of the soil, and to learn how to work hard and methodically for long periods at a time. There is not the same physical effort needed today, it is all done by machine, but the application and perseverance are just as essential as they ever were.

When I started real farming, two horses would plough about an acre in a day. My present tractor driver does 15 acres in the same time; he sits in an air-conditioned cab, listening to the radio. I used to stumble along in the furrow. But tractors don't steer themselves as horses very often used to walking up the furrow. Any inattention by the tractor driver will lead to deviation and, as it goes much faster than horses, the deviation will cause quite a big wobble before it is corrected.

There is nothing difficult in learning how to drive a tractor and use the various implements. But the operator needs to develop an understanding of when the land is fit to work, and when it should be left alone. Horse work was good training, because your feet were on the ground.

When following the plough, harrow or rolls, you soon learnt the difference between the possible and the impossible, when trying to make a tilth.

This just cannot be taught in a lecture room, or out of a book, or even on a college farm. It can only come through being intimately connected with a farming atmosphere. My sons all learnt these elements unconsciously, through just living amongst the sights and sounds of farming. The methods they picked up may not have been the most modern, but they were based on practice which was financially successful. One of them went to university and another to college and although they enjoyed and appreciated the experience, they told me that what struck them most about it, was the distance between the teaching and the practical realities of farming.

This is not to run down further education as such, but to underline the fact that, for farming anyway, academic courses are useless unless the student has a sound knowledge of practical farming to begin with. Once he has that, either from his own farm or on someone else's, it does not matter how he educates himself further. Some of the best farmers I know have graduated in arts or sciences, but they either had a farming background or sufficient capital to start half way up the farming ladder and pay for the mistakes of ignorance.

Leeds University and P.G. Holder certainly came between me and grassroots farming but in retrospect I don't think they really held me back for too long. At Leeds in particular I made new friends and entered a completely different social class. Many of the students were already men, who had worked themselves up from the coal face or the mill through night school and scholarships to undergraduate status. A degree for them meant that they would escape manual work, and either qualify as mine or mill managers or become teachers, a very sought-after profession. The West Riding was a hard place in the late

1920s and they fully appreciated the opportunity they had been given to get out of the rut.

They had very little money. But then the cost of living was low. A supper of fish and chips and tea cost 6d (2½p) and you could ride for miles on the buses or trams for a penny or two. There were two theatres and a music hall in the town: a gallery seat was, if I remember right, a tanner (6d) too. I can't say that these had any definite cultural impact on me, but I did sit through a lot of Wagner, Gilbert and Sullivan, and various plays by Shakespeare and others during the time.

There was very little in the way of student protest. We did get very worked up about something. I forget what, and several hundred of us marched, escorted by the police to Leeds City Square. There somehow or other we found ourselves queuing into a big cinema and having to pay 1s (5p) for the privilege. The film was dreadful and one or two bold spirits began to catcall and try to sing. They were immediately removed by a couple of enormous commissionaires and told by the police sergeant outside to go home quietly. They did.

There was very little feminine company, the textile department being particularly barren of this delight. The arts and medical students had first pick in those faculties where the balance of the sexes was rather more even. We were extremely jealous of this, and spent a good deal of time trying to muscle in on them, but without much luck. This sort of enforced morality and frustration did us no lasting harm, I am sure. Most young men of my own, and perhaps even the present generation, would boast of their conquests and talk endlessly of sex in a way which I discovered eventually to be academic, like the farming which was taught completely without practical knowledge. Our morality was based on negative standards. We had little money with which to patronise the tarts who plied for hire. Even if we had, the danger of disease was enough to keep us clear of them.

Then if one did pick up a nice girl – i.e. one who would come with you for the pleasure of your company – she was, if she was really nice, extremely careful because sexual education of those days being what it was, the knowledge of birth control was elementary. No one wanted to start a baby; the girl for obvious reasons, and the boy because of the fuss of the affiliation order or the forced marriage. Events of this sort had surfaced in my home town and the shamed participants were hurriedly exiled to distant parts. Lives we understood had been ruined.

This must all look very silly in these days of universal promiscuity, the pill, penicillin and abortion on demand. I am sure I would make the fullest use of these advantages were I to be young today. But although a great deal of frustration could be avoided, I doubt if, on the whole, I would be any happier.

At Leeds the other great thing was that I was clear of school and the home influence. Living conditions in my hostel were, I have learnt lately, considered fairly spartan, even by the standards of those days; but compared with school they were marvellous. A bedsitter to myself, three good meals a day and the freedom to walk in and out when I pleased seemed heaven to me. There was no criticism of my behaviour or of the company I kept, and best of all I could choose it.

The great advantage I gained at Leeds was to begin to break down the class consciousness which I had absorbed throughout my upbringing and schooling. Almost all my friends were of what would have been called working-class origin. The sort of people I would never have been able to associate with at home, or if I had, certainly not the sort of people whom my family would have considered suitable friends.

I have never believed that all men are born equal in

character, ability and so on. There are many different levels of interest which tend to segregate those of like minds in later life into their own groups. There is no crime in choosing your own friends when you are old enough to do so; but to force children, as we were forced, into recognised stratas of society from their earliest days has been the cause of much of the industrial trouble we are suffering today.

It is worth noting that the countries which are said to be outpacing us: Germany, France and the U.S.A., all have universal education through a state system. It is a system very much based on merit for promotion to higher levels; and I am sure that it is the basis for the much better communication between employers and employed which exists there. I cannot claim that at Leeds I was really associating with the real working class, but with men who by their efforts were escaping from it, so they were exceptions themselves. But it was a start.

My time in Yorkshire finished up with some months in the Airedale Combing Company Mill near Bradford. This was a survivor of the dark satanic mills and, at the time I was there, was enjoying the boom which made for considerable cheerfulness among the workforce. My particular job was learning wool-sorting, taking the fleeces out of the bales in which they had come from Australia, New Zealand and other places and throwing them to the sorters' tables.

These sorters were the aristocrats of the trade; grave, rather ponderous men with astonishing quickness of hand and eye. As soon as the fleece was spread out before them, they divided it into the different counts, threw the result into separate bins and pointed to the table so that I and the other apprentices should throw out another fleece. They worked very fast and if I was clumsy unrolling a fleece or slow in breaking open another bale, my own particular sorter – Joe, the head-man – would talk seriously to his neighbour, never to me directly, about the useless people God was creating these days.

Joe's neighbouring sorter, a squat and ugly man, was a great womaniser, or so he said. Like the rest of the sorters, he would titivate our ears with tales of his conquests step by step. At the dinner hour he would disappear into the wool-store, a dark place at the best of times, in search of the mill girls whom he claimed were lurking there looking for men with whom to satisfy their randy passions. The other sorters goaded us to do the same; but the sort of woman working in our part of the mill looked to us to be dangerously rugged and showed no sign of being attracted by our hairless faces.

I also became acquainted with restrictive practices. There was only so much wool to be sorted each day and the men on piece work shared it between them. There was one very quick sorter – a Scotsman – who, when he had achieved the Union's permitted stint, would read a book or sleep on a wool-bale until the whistle blew.

I can't say I learnt very much about wool-sorting, but I got the general drift of it and I also became quite good at sweeping up the floor, and wheeling heavy skips about. Finally I got the idea that a day's work meant eight hours solid with a break for lunch, and that it was better to work than to idle, or Joe would come down on me like a ton of bricks. If I had stayed in the wool business, I should have had five years of this, gradually going through the mill in every sense. But I was still set on farming, and persuaded my parents to let me go to New Zealand at the end of the summer. My departure from the mill was coincident with a slump in the wool trade which reached mammoth proportions within a year or two, so I was well out of it.

3. Emigration

My choice of New Zealand for farming was arbitrary and based on a complete ignorance of the facts about world or British farming. There was, during the 1920s, a great deal of propaganda to the effect that farming was better in the Dominions than at home. Neither my parents or any of our friends knew what conditions were like in farming there: but emigration was encouraged, and the papers carried advertisements of bronzed young men in khaki shorts with wide hats doing various things in sunny fields. There was a belief that there were huge areas of land for the taking, and various overseas governments were running settlement schemes. Farming in England had been going gently downhill, and in any case was most expensive to get into.

Once the decision to emigrate had been taken – and it was one I favoured because it meant getting clean away from family and other associations – the choice was narrowed to three countries: Canada, Australia and New Zealand. Canada was out because three of my father's cousins had gone there at the turn of the century. Some of them were still there, and the accounts of their misfortunes, the appalling cold, and the dreadfully low prices turned us against that country. In point of fact, it was the only one where land could still be had for nothing and a start was still possible, but a hard one.

All we knew of Australia was that it had been settled by convicts and it was a rough place subject to drought. So New Zealand it had to be. It had the added advantage, according to my parents, that we had relatives there including a Bishop. But their advice when it was asked, was that farming was difficult too. There was at the time a scheme for public school settlement. Boys went out to be apprenticed on selected farms, mainly the big squatocracy, but I was determined to be independent.

I was singularly unfitted for colonial life. I had no manual skills at all. I could not even milk a cow, plough a furrow or shoe a horse. Nor could I lay bricks, concrete or a parquet floor. My administrative skills were negligible as I had no accountancy training or typing. But I could read, write and speak French and ride a horse to a certain extent and a motorcycle.

In fact I was quite good with motor bikes and cars. My hobby before I left England was motor-cycling and I could always do elementary maintenance on the internal combustion engine and would be a handy fellow to have on a farm where tractors were used. Only in New Zealand there were no tractors. As there were at the time of my arrival in New Zealand, some 30,000 unemployed, all having the same basic skills that I had, hanging about the towns; the most sensible course for me would have been to remain at home and learn a trade. But no one told me, and in any case I don't suppose I would have listened if they had.

My father paid my passage out and gave me £15 for the journey. I had a few pounds in my bank account, and a motor bike which I remember cost £9 to ship to Dunedin. My father also gave me £500, half the cost of sending me to Cambridge which he thought would be enough to give me a start in farming should that become necessary. That £500 remained in sundry bank accounts earning about 1 per cent interest for the next few years. I was determined to make good and not to become a burden on my parents or anyone else.

By an extraordinary exercise of parsimony, I worked my way round the world on my earnings over the next four years. In a way, I failed to take the best advantage of the opportunity. If the main anxiety in any new port of call is the urge to get a job or a cheap lodging or a move on, there is not much time for sightseeing. So it was a pretty constricted experience.

I travelled to New Zealand on a cargo ship, the *Port Sydney*, which carried about a dozen passengers. There were no girls of my age aboard, a fact which I am sure that my mother had ascertained before the passage was booked, so propinquity didn't have much chance of getting to work. I am sure I missed a lot by not travelling on a passenger liner with its many distractions, but I was a very priggish youth, and almost revelled in my enforced asceticism. Nor did I drink or smoke at that time.

Sea passages are boring and the view of a ship's wake is much the same whichever ocean you are on. My cabin companion was a resident magistrate under the Australian Government in New Guinea going back from leave. He suffered badly from malaria and his teeth chattering during the rigors which reached a crescendo during the attacks worried me. They worried me still more when I suffered the same illness some years later. Fortunately I have shaken it off, but I am full of admiration for anyone like him who went back to a miserable climate to earn a living for a wife and family whom he only saw on his triennial home-leave.

I spent most of my time with the apprentices and junior officers, and to occupy myself had the Captain's permission to do the apprentices' work with them. Apprentices had a special position in those days. They earned no pay and, in many cases, their parents paid a premium. In return they did almost every job on the ship – many of which the unionised crew would not do – and in return they were supposed to be taught navigation and seamanship for up to four years.

The work consisted mainly of chipping paint and then

repainting, with a little steering and attendance on the officers on the bridge as well. During the voyage some of the crew broke into the hold and stole some of the cargo, principally whisky; so the apprentices, being of officer material, were put on cargo watch – four hours on and four hours off – to prevent further loss. I shared this duty too and had to suffer with them the catcalls and mockery of the sailors, who looked on us as the bosses' stooges or worse.

l had considered briefly the merchant navy in my youth, but after having made this and other voyages I was pleased that I had not taken it up. The boredom of long sea voyages and the bad pay and conditions in those prewar days made it a pretty dead-end occupation. A cargo ship was a carbon copy of the caste system which operated throughout Britain. There were the deck officers, engineers and apprentices living in the main deck cabins. The N.C.O.'s, as it were – quartermasters, bosun and lamp trimmer – lived in the poop deck-house. The rest of the crew lived in the forecastle, the approach to which was washed by seas whenever it was rough and where the conditions were deplorable. Of the crew the seamen ranked above the firemen who fed the boilers and who were referred to as scum by all the officers. No wonder they broke into the cargo!

We stopped to coal at both Port Said and Perim and I saw for the first time long lines of natives, our word for them, carrying the coal aboard in little baskets. Quite a few baskets spilt their loads as they ran up the gangways, and the next morning men were diving for the lumps and bringing them coughing and spluttering to the surface for their friends to take away. Diving like this must have been hard work as the water would have been around 30 feet deep.

I went ashore at Port Said, but can remember little of the city. Warned by the steward of the pickpockets and other harpies who frequented the place, I took very little money ashore and kept well in sight of the harbour in case I should be waylaid, drugged and robbed. An eye was kept open for the French women who were reputed to seduce their customers by exposing their delights in carriages into which they would entice them. This was the main burden of a lecture to the apprentices and myself by the First Officer who threatened that anyone who succumbed would almost certainly be infected by clap or the pox, and would be left to rot in hospital at the next port of call. However none showed themselves.

I spent four days in Melbourne waiting for a passage to New Zealand. I had introductions to people there, but I had decided before I left that I would keep these for use in absolute emergencies as I would be better off making my own way. This was another mistake because without sacrificing my independence, I could well have had a much better chance of seeing the country and appreciating its inhabitants. As it was I attended the Melbourne Cup, of which I remember nothing of the horses but several fights in the public enclosure were quite spectacular. I also developed a taste for oysters and crayfish which I have never lost.

The *Manuka* in which I sailed to New Zealand was an old passenger ship and the complement included a circus, a theatrical company, and some quite attractive girls at whom I began to look with interest during the first dinner before we cast off. But once out of Port Phillip heads the ship began to pitch and roll alarmingly and kept it up for the four days of the voyage. I am a good sailor and kept on my feet, but no females under sixty seemed to be able to, so there was no romance. I never saw them again.

I landed at the Bluff, New Zealand's southernmost port. At that time, it was no more than a village with a quay and a corrugated iron freezing works by the shore. I spent the night in Invercargill, a bare bleak place with enormously wide streets. It is said to be the southernmost city in the world and on a wet Sunday afternoon it seems to be as far as anyone could be from anywhere.

The southern part of the South Island was originally settled by Scotsmen. Dunedin they looked upon as a sort of Edinburgh, and many of the towns and villages have Scottish names. Even the all-pervading Australian cockney, which is the basis of Antipodean English, was mixed with a Scottish accent. And Scottish values were very strong. Invercargill was dry at the time of my arrival and for many years afterwards; but Wallacetown, six miles out and in another county, was wet. Taxi drivers charged 1s (5p) a head to take the drinkers out but nothing to bring them back and ran a profitable shuttle service for many years until eventually Invercargill voted itself wet.

The Wallacetown politicians were Catholic Irish, and this element was very suspect all through Otago and Southland, as being the source of all the crime, which was mainly sheep-stealing, drinking and fighting. They were also reputed to be idle and improvident, in contrast to the sober Scots who ran them down continually. This bias was based on religion and went right back to Mary Queen of Scots or even further; it was a very real part of their lives.

The next day I travelled by train to Dunedin through what I thought to be a very disappointing countryside. Instead of the wide open spaces which I expected, most of the country through which the railway passed was fairly closely settled with quite small fields, many hedges of gorse or hawthorn and villages every few miles or small towns. There were hills and mountains on the horizon on each side, many of them were thickly coated with gorse and scrub. The houses were almost entirely wooden bungalows, creamwashed with red corrugated iron roofs. The roads were unsealed and the occasional car threw up great clouds of dust. People came to meet the train at many of the stations, and quite a few were on horseback or in traps. I noticed that most of the cars were very old models indeed.

Quite a lot of the land was in plough, and there were horse teams working. Not the slow and massive Shires which were then common in the south of England but the much more active Clydesdale. I noticed that, instead of walking behind the team as did the English carter, the driver rode on the plough or cultivator, or even on a horse behind the team.

The livestock in the fields were much the same as at home, but the overall impression was untidy. The hedges and fences were not kept up and the houses and buildings needed paint. New Zealand farming was just beginning to feel the effects of the post-war depression which was to deepen dramatically in the years to come.

I had chosen the very worst moment possible for my arrival. Farm prices had peaked in about 1921, and then had slumped steadily since. This was brought home to me on the first day I reached Dunedin. I had been advised to contact a Mr Draper, the manager of one of the pastoral companies which controlled much of the farming. Mr Draper took me first to the local market where fatstock for the town were sold. This he told me was the high price sector. There I saw cattle and sheep sold at little more than half the price that similar stock had made at Ludlow market two months before. This seemed extraordinary to me, but I reasoned that there must be some magic in the New Zealand environment which could make farm prices like this profitable.

That evening he took me round some farms about 20 miles south of Dunedin, good land and a lot of it dead-level and capable of growing good crops. In those days it was either dairying or arable.

'How much was this worth?' I asked him.

'Around the £100 an acre.'

And what about some very ordinary looking fields on

the foothills of the mountain which overlooked the plain.

'Between £60 and £70.' But he added 'Prices were depressed.'

Just two months before I had witnessed Holder's signature on a contract to buy a very nice farm in the Corve Dale in Shropshire at £45 an acre. From that moment I knew instinctively that it must be better to put my money, even my £500, into English farming where land was half the price and livestock double.

The reason for the high land prices in New Zealand is, as I discovered later, the streak in human nature that makes a man try and have a business himself instead of working for someone else. In Britain and in most other industrialised countries opportunities for self-employment abound. In New Zealand they were, and still are, mainly confined to farming. So there is an enormous pressure to get hold of land and farm it, which drives prices up to such an extent that farmers seldom get their heads above the load of debt which they had to take on to start with. It is as simple as that.

This has led to a situation in New Zealand where those taking up farming still have to pay some of the highest land prices in the world, and at the same time accept some of the lowest prices going. There are advantages in the New Zealand climate, particularly for grass growing, but in my opinion these in no way compensate for the very high cost of land.

I did not realise all this at the time myself, but I was quite shocked that the imaginings that I had had of land for the taking which had been one of the motives for coming to New Zealand no longer existed. It was, of course, my own fault for not having made proper enquiries before I left. But then I subsequently met men who had migrated under the various schemes of the time, who found that, while not being exactly misled, had certainly not been warned.

Emigrants to Australia of the period had much the same experience, often meeting downright hostility from the locals when competing for jobs. The fault lay with the local politicians who understandably enough were seeking to replenish the stock of lives lost in the war, and to build a white population to counter the discernible threat from the coloured races in the East. It was felt at the time that Australia and New Zealand were much too thinly populated to support an industrial base and that increasing numbers attracted on whatever pretext were better than none at all. Since World War II, immigration has been extraordinarily successful, especially in Australia, but it has been planned and developed with care and at great expense. Fifty years ago, the new immigrant was simply pitched on the beach to fend for himself unless he had friends.

These I did have and made more as time went on. I have often wondered how I would have got on without their help and had been just one of a crowd of 'bastard poms'. Not so well initially perhaps, but in the end friends were perhaps an inhibiting factor. Certainly I got jobs with their help, but then if I didn't like the job or saw a better opportunity I did not move on, simply because I did not wish to disappoint those who had gone out of their way to help me.

4. Glencairn

My first attempts at getting a job were pretty disastrous. Mr Draper said that the manager of a very good sheep stud on the coast north of Dunedin could do with a young man and he would take me to see him. The stud was one of the best in New Zealand. The manager was a most capable individual I am sure, and I would have learnt a great deal there. I was shown a bunkhouse and an office where I was told I would spend most of the next year.

This was appalling. Here I had come 12,000 miles across the oceans to do what I could quite easily have done at home. I looked through the books and at the old typewriter and thought of my failure even to get a pass in arithmetic. But I need not have worried. Could I do double-entry bookkeeping? Keep a journal? Measure a paddock? Could I type? No, I answered truthfully. Could I then build a wall, hang a five-barred gate or butcher a sheep? Again no. Well what the bloody hell can you do? It was, in the modern phrase, a good question.

So Mr Draper drove me back to Dunedin and on the way gave me the benefit of some very good advice. Never admit in New Zealand or in Australia, he said that you cannot do a job as long as you have the haziest notion of what it is all about. If you have no skill at all, just say, 'I'll give it a go.' He had, he told me reached his present position by means of this formula, and now he asked all the questions.

'Never in my hearing,' he added, 'will you ever say "No" again.'

A day or two later we went to one of the local shows

near a village called Outram where I was introduced to a farmer who said he could do with a boy for a while. Mr Draper did most of the talking, while Mr Park who was heavily in debt to his firm simply nodded. I acquiesced in the transaction eager to avoid letting Mr Draper down again.

A few days later Draper drove me out to the farm and along the road we met Mr Park mending a fence—but a very different Mr Park. I expected a bronzed and sunburnt character in rolled sleeves. Instead he obviously hadn't shaved since I had met him first, his clothes were throwouts from his town relations, and his boots had no laces. I felt very overdressed and went on feeling that way until I stopped shaving.

Shaving for some reason was important in my family. My father's last words as I went on board ship was to remember to shave every day, it was, he said, the mark of the Englishman the world over. He had never been further than France of course. So I shaved religiously in cold water every morning for about a month, while Park shaved about once a fortnight.

After a month he asked me if I liked being there.

'Oh yes,' I said.

'Well, if you want to stay you have to stop shaving, the wife is making comparisons.'

So I stopped.

As I came to know him better, I realised that Park was not a good farmer. He lacked the basic instincts of a businessman and had no idea of the economics of any farming operation. He had enlisted at 15 and spent most of the war in Egypt and in France where he had been badly gassed. On his return, he had been shepherding and rabbit-catching on some of the big mountain sheep stations, had saved some money, and had taken advantage of the New Zealand Government's very generous scheme for the resettlement of returned soldiers.

The farm was one which had broken a number of hearts since it had first been settled in the 1870s. It covered about 5,000 acres, but only about a quarter of this was level land on which a team could be driven. The remainder was composed of steep gullies running up to central ridges. The Taeri river was one boundary which was heavily bushed and the vegetation in the other gullies was either native bush or manuka scrub. This plant, rather similar to juniper, took over everywhere as soon as the original bush was cleared. Much of the work I did was concerned either with burning it or chopping it down by hand.

Originally the land had been ploughed and oats grown for the horses of Dunedin. No ploughing, however, had been done since the war and the pasture – which is called Browntop – was of very little use for anything. The total carrying capacity was 1,000 sheep – of which about half were ewes – and some 35 cattle.

Park was trying to increase the carrying capacity of the farm by ploughing and reseeding. This was an expensive job even at the prices of those days. He hired a team of horses and I had my first experience of ploughing and working land. I must say one of the pleasures of farming in the twentieth century has been the disappearance of the horse. They are stupid animals at best, and require almost as much effort to look after as the work they perform generates. I had to get up at 5 a.m. to feed and groom them. Then I did the other chores, had breakfast and then drove them out to plough or harrow. In the evening the process would be reversed.

Neither of us knew much about arable work, but in the end we ploughed up and worked down some 30 acres which we sowed with ryegrass and turnips. The seeds germinated, the plants appeared, then they withered and died. No one knew why this had happened because the fertiliser used had given excellent results in other parts of

New Zealand. He was beaten by the mineral deficiencies which made it impossible to grow clover or reasonable grasses. Research on this was in its infancy in those days, and it was many years before it was found that three or four ounces of molybdenum per acre was the catalyst which made the fertilisers, seed and cultivations produce a reasonable pasture. Some years ago I revisited the farm; it had been divided into two farms and the total stocking was some 10,000 sheep. Poor Park, alas, had died even before he could have seen the results.

I had been mindful of Mr Draper's advice that I should never admit to be unable to do anything, so when the boss gave me a bucket and told me the cow was tied up ready to be milked in the barn I went off to find her. I can see her now – a big black Aberdeen Angus cow tied not only by the neck but leg-roped as well.

This leg-roping should have warned me, had I known anything of milking. But although I knew the theory, I had no idea of the practice. I knew it concerned squeezing the teats, so I pulled the stool up, put the bucket under the udder and started to squeeze the two nearest teats. As soon as I did, the cow started to twist and turn and almost hung herself on the two ropes. If her leg had not been roped back, she would have kicked me out of the shed. As it was she almost fell on me.

I kept doggedly on but after about a quarter of an hour I lost all power in my wrists and had no more than a couple of inches in the bucket. I was sure she had some more in her but was equally sure she was not going to let me take it or at least that my wrists were not. On my way up the house was a little stream, or creek as they call them in New Zealand. It was the work of a moment to add another inch to the level of the bucket.

'You did well,' said Mr Park when I showed him the result of my labours. 'The last boy I had here just did not know how to handle her and she nearly killed him.' After that he left the milking to me and within a few days I felt I could safely manage without the creek water. No one noticed the difference, except the baby grizzled a bit.

A day or two later I was given a knife and a bucket and shown a sheep in the yard. 'Kill and dress it,' was the order.

'I am afraid,' I said, 'you will have to show me how to do it. We don't kill sheep on farms in England.'

This set Park off on a diatribe about all the useless Englishmen he had met during his time in the army. I grew accustomed to this after a while, a public school education allows abuse to flow off your back without hurting.

In the months that followed I provoked many outbursts of this sort. The main job for the first few days was mustering the sheep for shearing. As I had no dog Park sold me one for £6, or 6 weeks wages. He was a big clumsy animal rather like an Alsatian called Dick and Park had bought him off a drover. He had little sense and nor had I. I had never worked dogs before: he was also very disobedient. Another disadvantage was that, being in the southern hemisphere, I had not much idea of direction because the sun was to the north. This was important when I got out on the run, because there were few recognisable landmarks: all manuka scrub or bare pasture looks much the same.

In parts of the run, there were big acreages of burnt manuka which the sheep used to graze among. You could ride through them, but visibility was no more than twenty yards. If I saw a few sheep, I tried to drive them in the general direction of a fence down which I reasoned they would go until they found a gate. But the fences were so bad that the sheep just streamed through them, often pursued by Dick who barked fiercely most of the time. Sometimes he brought them right back past me, sometimes he lost his temper and bit them.

For the first few days I was lost for most of the time,

and can only conclude that the noise Dick and I made moved them in the right direction. We eventually got them into a block of fairly well fenced land near the farmhouse and then to the yards by the shed. Then we set out to remuster the farm because we only had about half the number turned out some months before.

Some of these had gone on to neighbours' farms, and one of my jobs was to ride over and bring back the stragglers. Dick was quite good on the roads with a small mob and I did not make such a muddle of this. One of the neighbours had quite a reputation for appropriating other people's sheep. He could not keep the ewes which carried Park's earmarks, but I remember that he only allowed me to take away one lamb with about 16 ewes.

'Was this all they had?' I asked.

'Oh yes.' he said, 'They were barren or the hawks had had the lambs.' As we walked home, I noticed all the ewes were leaking milk which showed that they must have been suckling the day before.

This character was once found skinning a sheep at the side of the road by a constable who had ridden up silently on the grass verge. He obviously saw the Broad Arrow Government mark on the hooves of the constable's horse because, without looking up, he sliced off the ears and threw them to his dog who promptly gulped them down. Then he looked up and said, 'Awfully fond of ears this little dog of mine.'

I had to report to the constable on my arrival because I was liable for conscription and I was duly attested to the Otago Mounted Rifles. As I lived about 20 miles from the drill hall and had been in the O.T.C. at school, I was excused drills, but would have to attend the annual camp for three weeks. I also had to take a firing test which I passed.

Although my knowledge of infantry training was quite good, I had no idea of mounted work. I taught my horse, which was the one I would take to camp, to stand still while I fired a rifle from the saddle but could not persuade it to lie down so that I could fire over its back. This manoeuvre, which was of doubtful military value, was always illustrated in the photographs of the annual camp. Park and I tried hard but to no avail. I drew my uniform and gear and was all set to ride off to camp when Park succumbed to pneumonia. As there was no one else on the farm, I had to stay. By the time the next camp came round, conscription had been called off.

Today New Zealand is infested with opossums, and the roads are littered with their corpses, but 50 years ago they were scarce and their trapping was strictly controlled. On the other side of the Taeri river was a big expanse of native bush where one could see all the traces of their presence. To increase his income Park had taken a block in Canterbury for the trapping season but, not content with that, he had trapped before the season opened in the neighbouring bush while the man holding the rights had been respecting the season. He caught quite a few but, when he went up to Canterbury, he found that someone else had been there before him.

As a returned soldier, Park had received a grant and special consideration for a loan from a pastoral company. As a condition of the acceptance of the loan he had to buy and sell all his stock and requirements through the company or stock agent as it was called. Few New Zealand farmers in those days, or even today, go to market or ring the changes between different suppliers or customers. In fact it was, and still is, considered rather bad form to do so.

The New Zealand farmer was a farmer first; generally speaking he stayed at home, worked his farm and left the business to the agents. If he wanted more sheep he asked the agents to find them, and if he wanted to sell some he did the same. All his wool was sold for him by an agent's auction. If he was deeply in debt he was put on a budget and his food, clothes and other necessaries supplied.

This system would not have suited me. There were far too many opportunities for exploitation. In fairness it must be said that, by and large, it worked well and many men got a start who would never have got off the ground without it. The weakness of the system has been that New Zealand farmers have never developed into the sort of farming businessmen able to run the co-operatives and marketing boards as in Britain. This I think has been responsible for the poor showing that New Zealand marketing organisations have made in disposing of their produce.

However at that time these considerations did not enter my mind. My job was to work and earn the £1.00 a week and my keep that I was paid. Apart from getting the sheep in for shearing, I saw little of them. A shepherd did not look after his flock in the English way at all. If it was fine weather there was fencing; if the materials ran out, there was always plenty of manuka scrub to cut down. I learnt to use an axe and, although I say it myself, I eventually became quite good with it.

The manuka was I suppose about 20 or so years old and about two to three inches thick. Properly swung, a sharpened axe could get through it with one blow. I used to trim off the top branches and 'snig' (drag) the stems back for firewood with a horse. It was our only fuel and the kitchen stove used to burn two or three hundredweight a day. A snig is a loop of chain and this attached to a horse collar was the only power on the farm apart from ourselves.

If the weather was wet, we clipped dags. That is we cut the wool off the manure clots which the shearing gang had taken off the fleeces. Every woolshed I ever visited had a great pile of these waiting for a wet day. It made quite a social occasion, sitting round the heap of dags with the boss and almost anyone who came to call or was working temporarily near at hand, such as a roadman or rabbiter. I learnt all the gossip of the neighbourhood in more than adequate measure. Even the stock agent would pull on his overalls and lend a hand for a while.

After a while, I was trusted to look after the farm by myself while Park and his wife went away. Before going Mrs Park, who was a very good and economical housewife which was essential in the circumstances of the time, had pickled a sheep. For this she had used a recipe handed down through her family from Scotland.

'Just make stews John,' she advised me on leaving, 'pickled mutton won't roast or fry.'

So I duly attempted to make the stews according to her instructions.

Somewhere in the passage from Fife to Otago a vital ingredient in this recipe must have been lost because the meat had gone bad, or at least it tasted bad. I was young and hungry and had been to a public school, so I ate some of it each day, but mainly lived on porridge, milk and cream potatoes and bread and butter. I made the butter myself and baked scones or soda bread. One day a friend of Park's appeared and said he was going to stay for a while as things were bad in the City. I strongly suspect he was in some trouble there. For the first supper I served up the stew. After about two mouthfuls he exploded.

Didn't I know the meat was bad? We were bound to be poisoned here miles from any help, he had seen men in the army die from far less rotten meat than this. I explained the situation which was to save killing a sheep just for myself I had been left this pickled meat.

'Never mind,' he said, 'Are there any killers in the yards?'

There were some wethers which we killed for the house: in a very short time I had one killed and dressed and some very tough but fresh chops were in the frying pan.

I emptied the pickle barrel and gave the meat to the dogs. They were only fed every other day and were usually pretty ravenous, but they had about two sniffs and howled for something else. Eventually I carted the remains into the scrub where they remained intact for a long time.

Once back on fresh meat I began to pick up my appetite. I was working a 12-hour day and cooking for my guest and myself morning and evening. He spent most of his time in bed and said he was still suffering from his war wounds. We had quite a good garden but, because I had been forced to eat green vegetables all my young life, I simply fried up potatoes to eat with the chops and had no fruit or other supplement beyond porridge. After about three weeks I developed horrible sores on my arms and chest. I consulted my guest.

'That looks bad,' he said, 'When did you last have a woman? I saw a chap like that in Port Said; you can't be too careful even in Dunedin.'

I said I hadn't had anything to do with women.

'You must have sat on a lavatory seat then. Permanganate of potash is the thing, soak your prick in it every night.'

I hadn't sat on lavatory seats which could have carried the infection for months, so I went to the doctor. I tied my horse at his gate and walked into the surgery. He examined my sores.

'Been on the frying pan my boy?' I admitted that I had.

'All right,' he said, 'Pick four cabbages from my garden, buy a case of apples at the store, and eat them raw or cooked. If at the end of that time, you still have the sores we will have a good deal to worry about.'

I did as I was told and the sores were gone in a week. I had suffered from scurvy, and ever since I have had a passion for fruit and vegetables.

While Park was away, I was supposed to look round

one lot of young sheep which were on a 1,000 acre block on the banks of the Taeri, steep and much of it covered with native bush. One of the features of this was a small thorn-like briar. It was called lawyer locally, because the thorns went both ways! As the winter progressed, the sheep forced their way further and further into the bushes to get at the green leaves. In doing so, they would get caught up and the more they struggled, the more they were entangled. Once this had happened they became easy prey for the hawks who would start by picking their eyes out and then feed off them.

Checking them out was quite hard work, and to do it thoroughly would take anything up to a couple of days, because it meant climbing up and down the steep slopes. You could spot a sheep in trouble if you saw hawks hovering over one in the process of being caught up. I was supposed to do this every week and I saved on average half a dozen after each search. When Park returned he decided that I had not cut enough scrub as I had only been cutting for four days a week instead of six. I showed him how it was impossible to check the sheep thoroughly in less than two days. I also pointed out that for the price of £1.00 in cash and my keep I was saving him £9 worth of sheep.

'That is not the way I see it,' he replied.

This was a salutary lesson to me. I realised then that many hardworking New Zealand farmers did not get their priorities right and that practical skill alone won't get anyone very far.

But life was not all work and isolation. Although I was alone for quite a bit of the time, I did not feel lonely. After all the chores and the work I used to hit my pillow at about 9 o'clock and sleep like a log until an alarm went at 6.30. Except once: I was sleeping in a room called the whare, pronounced warry, at the end of the big woolshed. I woke with a most peculiar feeling that something was going to happen. Then I heard a noise like a tube-train passing under a house, the whole building rattled over my head and a lot of soot came down the chimney, then all went quiet. A day or two later, when I went to town to get the mail, I heard there had been an earthquake.

On Saturdays there were dances in the nearest village; and on Sundays, if there was nothing to do, I could go visiting. The Saturday dance was to the music of a brass band, no discotheque in those days and the foxtrot had hardly been heard of by the dancers. There were some quite nice girls around, but very sensibly they weren't going to get involved with a young Englishman with no apparent prospects, when there were farmers' sons around with farms to go to, or else they were set to gravitate to the towns.

Still I did visit one or two in their homes. The only way to get anywhere near them was to offer to do the washing-up with them after tea; no one ate supper or dinner there. In this way I became a very good washer-upper but that was all: I was too shy or too careful. Shotgun marriages were much the rule in those days, and I did not want to start something I would not be able to escape from.

Nearly thirty years afterwards I came back to this place and met some of the girls I used to know; by this time they were well on the way to being grandmothers. Looking at them I felt a sense of relief that nothing had happened to link me with them for life, only to see them appraising me with the same look on their faces. There is something to be said for growing old together.

5. Shearing and Threshing

My next place was much more like the sheep farm of the travel books. It was about 18,000 acres – mostly very high and unimprovable land – carrying about 6,000 sheep altogether. There was no attempt here to do any ploughing or improvement. It had all been fenced some thirty or forty years before and much of the work the shepherds had to do – there were two of us – was concerned with maintaining fences. Here again, as in most of New Zealand, there was very little sheep work as we used to do in England.

I was given charge of a lambing flock of about 1,000 ewes running on some warm slopes about 6 miles from the homestead. I used to leave as soon as it was light after having milked the cows, taken in wood for the house and helped cook the breakfast. The block was of about 2,000 acres, well broken up with gullies intermingled with some fairly dangerous bogs in which a horse could get stuck. There was no bush or scrub which was a blessing. By riding slowly round and using a pair of field glasses, I could cover a fair bit of the area and see most of the sheep.

There was little chance of saving any lambs, but I occasionally found a sheep which was trying to lamb and which I could assist. There was also the odd one on her back to which, with luck, I could get before she either died or the crows or hawks picked her eyes out. These birds were numerous and could spot a ewe in trouble a long time before I could. The hawks were quite tame and used to hover over me as I rode. I shot quite a few of them off the saddle as they were poised overhead. They caused enormous losses and would sometimes attack the lambs as they were being born.

There was an old homestead on the block and I suggested to the boss, a fairly stolid Scotsman, that I should camp out there for the lambing and organise the job as we do at home. That is to keep the lambing ewes in a smaller paddock so that they could be watched. I knew the lamb price was bad, about £1 a head; but if I could only save five a week I would save my wage and my keep and still show 100 per cent profit. But he did not think it worth while and the same attitude prevails today.

Every year now I employ one or two young New Zealanders to help my shepherd. They are first-class workers – farmers' sons and well used to sheep at home – but most cannot lamb a ewe when they come and have no idea of the tricks needed to persuade an unwilling ewe to take to a lamb. Part of this attitude stems from the low prices of sheep in New Zealand and the very large numbers a farmer has to keep in order to get a living at all.

It was on this farm that I first learnt to shear. It came about this way. The shearing gang was having its Sunday rest and one of the shearers, a neighbouring farmer's son, began to bait me and challenge me to a fight. He had some pretensions of being a boxer. I had done some boxing at school but after an early success had met a better man and then retired.

So when Morry, as he was called, shaped up and started to land a few taps on my head, I picked up a pair of shears with which to defend myself. Hand-shears are nasty weapons with blades some six or seven inches long, and very, very sharp. Morry took one look and mistook my defence for aggression. He backed away and, as I followed him, grabbed his bag from his bunk, shot outside and on to his horse, and away. He did not return on

Monday morning and the boss, who had heard what had gone on, simply said:

'Well, John, as you have got rid of him, you must take his place.'

The first sheep I shore was a dead one which had succumbed during the night; the head shearer demonstrated how it should be done, and let me start. Shearing follows a set pattern and one of the strokes or blows with the blade is to run up the throat under the wool to open the fleece. As I was laboriously doing this, I found the going hard; in the end I had cut the corpse's throat.

I never actually killed a sheep while shearing, but I certainly made some acutely uncomfortable. The sheep were a merino cross called the halfbred, and the skin was highly wrinkled. Also the wool was very sandy from the banks against which the sheep used to rub when they felt the urge to do so, probably because of the irritation of ticks. In addition the wool was very dry as the sheep were in fairly low condition. It all added up to a blueprint for hard shearing. On this run, and on many of the high country places, shearing machines were not used in those days as they got too close to the skin and made the sheep very vulnerable to bad weather. With hand-shears it was difficult to take off the last half inch.

Shearing is hard work, not only have the shears to be closed for every blow but have to be driven through the fleece as well. The sheep has to be held and turned in a regular sequence while the human body, mine, has to be kept in a permanently stooped position, thus putting considerable strain on a whole host of muscles singularly unused to such exercise.

Shearing was run to a timetable and the day was divided into six runs of about one and a half hours each, starting at 6 a.m. It is hungry work and snacks or whole meals came between each run. There was a start at 5.30 with morning tea, scones and cake. Breakfast was porridge, chops, etc., at 7.30 a.m. Morning tea again came between the two prelunch runs. There was a substantial midday meal, afternoon tea, then a proper cooked tea with three courses and we finished up with supper at 9 p.m.

I had used shears before to crutch or clear wool from the sheep's behinds, but although this taught me to handle shears it had not taught me to shear the whole sheep. The first day's work was hell. I got through about 40 sheep and had to endure a whole lot of chaff. The man I was replacing was doing about 150 and one man on the board aged 69 was doing rather more. Shearing exercises a whole lot of muscles that are never used at any other time, so at the end of the first day I was aching all over. The next day was agony, but I stuck it out and had worked up to about 60 when the main shearing finished and the gang was paid off.

'That's fine, John,' said the boss. 'Now we will muster the stragglers, and you and Rowan (the other shepherd) can shear them.'

I should explain that the stragglers are the sheep which have missed the first muster or which had been lost on the drive to the homestead or had to be fetched home from the neighbours.

Mustering made for a long day. I got up at 2.30 a.m., fed and groomed my horse and got in the two cows which I had to milk before breakfast. Then breakfast – porridge and chops – and a couple of hours ride in the darkness out to the block to be mustered. By the time we got there the dogs were tired, and by the time we had got the sheep gathered and back at the homestead it was often 7.00 or 8.00 p.m. Then I had to deal with my horse, milk the cows again, and have my supper and finally to bed at about 10 p.m. Often I had to start again the next day at 2.30 a.m. During the summer there were several weeks of this sort of work, and all for 30s (£1.50) a week and my keep.

Once the stragglers had been gathered they had to be shorn, and I was put on to shearing in competition with Rowan the other shepherd. We were running neck and neck when I thought I would steal a march on him by shearing one that had lost nearly all its wool. The boss, however, had an eagle eye, and stood over me while I carefully trimmed it all over – a much slower job than shearing a fully woolled animal. So I lost, but still managed more than 100 in a day which was not too bad with hand-shears and represented about my top in manual work.

Otherwise I learnt quite a lot about sheep-farming on the New Zealand pattern. The great thing was to keep costs down and on this place there was no farm work at all. The sheep were grazing the grass that grew naturally and the owners cashed in the natural increase and made their living from wool and the sale of surplus stock. There was no provision for winter feed; so if there was a lot of snow the sheep went hungry or died. We had one very bad snowfall in December, the midsummer month, and lost some lambs buried in the drifts.

I must say I enjoyed the life. The boss and his partner were very civilised and used to whistle when coming round a corner. Unlike Park they were people of substance, and because their land was unimprovable they had been able to keep the profits they had made in the wartime wool boom. The whare where I lived was connected to a power supply generated by a most complicated hydro scheme which heated the water among other things. The trouble was that the dynamo would trip out for some reason and then someone had to climb down the hill and adjust the trip some 500 feet below. This was hard work at the end of a day so, if the bosses were away, it used not to be done. But the pleasant time could not last. Once the work was over the boss said I could live on in the whare without pay, because prices were so low that he could not pay me. Meanwhile I was to get what work I could.

I went to work on an arable farm of about 300 acres on the Taeri plain to help the owner, Bill Marshall, get in his harvest. He was an Ulsterman with a marked dislike of the English. However a lot of his bad temper I think had been due to the fact that his farm had been flooded the year before and he had lost his harvest. It was a very cold farm and the whare where I lived had a mantelpiece but no fire. The alternative to going to bed, or going out visiting, was to spend the evening in the farmhouse playing whist with Marshall, his wife and an elderly Ulsterman living in. The result was that I had to sit with my back to the window in the draught while the other three were much nearer the fire. In the end he had to keep me on for three weeks after harvest until he had sold enough grain to have some cash in the bank to pay me.

Next I joined the threshing gang for a while. It was mainly composed of farmers' sons earning some cash as their fathers had precious little for them. In doing so, I got an interesting insight into the farms of the district and the people on them. They were mainly lowland Scots and their farms were probably some of the best farmed in New Zealand and run on the Scottish pattern of turnips and grain growing.

Most of the farmers were glad to see us, and their wives spared no pains in feeding us well in breaks in the long days. My job was loading. I drove an empty cart out to the rows of stooks, where the pitchers were waiting. One of these was the sheaf-tossing champion of the district and he used to send the sheaves whistling up to me. I had to build them into a sensible load that would not fall off on the drive back to the threshing drum.

The pitchers delighted in burying me in sheaves and, where the crop was thin and the stooks far apart, led the horse at a run so that the higher the load grew, the more it lurched on the uneven ground and building the load became almost impossible. If the worst happened and a corner of the load fell off, they hurled it back up again without mercy. To make matters worse the engine driver blew his whistle if I was delaying, at which I had to gather the reins and drive as best I could back to the thresher. Once there I had to pitch the sheaves from the cart to the man feeding the drum.

He was one of the most relaxed and fastest workers I have ever seen. I pitched the sheaves one by one and they had to fall exactly where he wanted them, he picked them up, cut the bond and fed them to the machine. However fast I pitched he was always clear and ready for the next one. He never said anything if a sheaf went awry, but he looked at me so contemptuously that I felt an utter worm. I had to endure a great deal of ribbing about 'useless Pommy bastards' from the rest of the gang during our meal breaks.

I was sorry when the threshing came to an end. It had upped my pay to about £2 per week, this for a twelve hour day with two hours off for meals. I have never worked so hard since. The experience did me no harm, and when talking about the degeneration of modern youth, as old people do, I like to give it as an example of what things were like when I was young. But it was no hardship, I was fit, well fed and enjoyed the experience.

That was the last spell of work I did in New Zealand. I travelled around the country visiting my relations in the North Island. While staying with an Aunt in Wellington, I heard of a load of sheep going to the Argentine, and that there was an opening for a stockman, for which the pay was a one-way ticket to Buenos Aires.

I left New Zealand with little regret for a variety of reasons. I had formed no close ties, and I found the world depression in prices which hit farming – New Zealand was the first of many countries – induced a feeling of despair among many young people. I wasn't the only young man looking for another opportunity by any means. The prospects were frankly awful, with lamb, wool and butter-fat prices at the lowest levels ever known. Few farmers could keep any men the year round. I could have worked for four or six months a year and spent the rest of the time among the unemployed.

Rural life was also very narrow. Some of the older men had been away to the war, but most of my generation had been no further than Dunedin or Invercargill. I wouldn't at that time claim any great intellectual superiority, but I had seen some of Europe and London, and realised that there was something else in life besides work, food and sleep for six or seven days a week, with the odd dance on Saturday night.

Not that I have ever minded constant work. But even in those days I thought it should lead somewhere beyond the daily bread and butter. I could see just no future for me as an independent farmer. Land in New Zealand even then was too dear to buy with any hopes of making a financial success. In any case I had no prospect of getting enough capital at the time even to pay the deposit on a farm.

l had discussed this with Bob Steele, my employer on the sheep farm, and he made the suggestion that perhaps at some future date I could save or raise a little more money and buy out his share of the big sheep run. I worked it out on the prices of those days and decided that, even if he gave me the run, it would be impossible to make it pay.

I must say, however, that I had nothing but kindness and hospitality from everyone I met. New Zealand was always a leader in the Welfare State business, and I knew that no one there would be allowed to starve. This is a great tribute to a country and one I came to appreciate more and more in my subsequent travels.

6. Beef in Argentina

When I got to the docks at Wellington where I was to join the *Port Pirie* on the morning of the sailing, I found an altercation going on between the shipper's agents and the veterinary inspectors. Many of the sheep turned out to have foot-rot, and rather more than half were not allowed to be shipped. The agent then came up to me and said that, as the work was going to be much reduced, I should have to pay a proportion of my fare. After some argument I gave him £12 and went aboard.

It turned out to be a cheap trip. As soon as we left Wellington harbour, we ran into head winds of great strength and the ship began a pitching motion which lasted for the next two weeks. The sheep were under the poop deckhouse in the stern which was rising and falling between twenty and thirty feet like an express lift. We chopped up the first day's ration of turnips, gave them hay and water, and they never ate a thing and seldom drank. After that we made friends with the First Officer who directed one of the stowaways, of whom there were three on board, to help us; so my companion John Vavasour and I became supervisors and lived the life of passengers. We had a cabin under the bridge, and messed with the officers and the few other passengers.

The voyage was quite an experience. The constant head winds threw up great seas which came crashing along the foredeck and against the bulkhead of our cabin with appalling violence. The din was continuous and we didn't seem to get much sleep. The crew lived in the forecastle and had to use lifelines all the time.

On one occasion the forward hatches gave way and the ship had to be slowed down for repairs to be made. She lay and wallowed in the troughs of the waves: it seemed that she was no bigger than a dinghy with seas the height of the Twickenham stands bearing down on her. On one occasion a freak cross-wave slopped up the side and over the boat deck, taking away two boats and doing quite a lot of damage; luckily there were no human casualties. I have never envied the round-the-world yachtmen.

This was before the days of radar and direction finding radio, and after about a fortnight the officers had to confess that they were not sure where they were as they had not seen the sun since we had left New Zealand and had been navigating on dead reckoning. So the ship was slowed, and lookouts were posted until land was sighted. Then we turned south and ran down the coast until they were able to fix a landmark and set a fresh course to go round the Horn. This was a bit of an anticlimax being the only calm day of the voyage, and I still have a snap I took of it from about five miles off. After this there was another week of rough, very cold weather before we reached Montevideo in Uruguay where we were to tranship to Buenos Aires.

I knew no Spanish but I had bought a phrase book before I left Wellington and had conscientiously studied during the journey. But pitch forked on to the dockside in Montevideo harbour on a Sunday evening, I found I knew less than nothing. However 'hotel' is a universal word and the first one we came to had a French-speaking manager. We had no money, only New Zealand banknotes which no one had ever seen. He trusted us, gave us a room, a good meal and the next day took us to a British bank. There the manager of the exchange department, a very washed-out sort of Briton, refused to change our money too.

'Never mind,' said the manager, 'I know someone else.'

Then he took us down a back street to a dingy little shop whose owner, some sort of middle Easterner I thought, made no bones about changing our notes at a good rate.

There is a moral in this tale. Fifty years ago British business was just coming to the end of a long period of absolute domination in the financial and trading life of Latin America, particularly in Argentina. The British were still trusted; *palabra Ingles* (word of an Englishman) still meant something. But the British had done too well: many of them never learnt the language properly, few of them mixed socially with even the rich Argentines. Their children were sent home to school. Those whose parents could not afford to do this knew that their children would be called *Portenos*, that they would speak English with an accent, and were in fact Argentinians.

Because British business was run by expatriates in the upper echelons, others were getting in beneath the upper crust and, like my Montevidean friends, beginning to take their business away. My time in Argentina coincided with the beginning of this change and I could see symptoms everywhere as I came to know the country better.

It was, of course, presumptuous for John and myself to think that jobs would be waiting for us when we got ashore. We knew a good deal about sheep work but we knew nothing of the language and we had few friends or connections except the agents to whom the sheep had been consigned. However, we had a market, not for our skills which were no better than those of the Argentine farm workers, but our honesty. It was considered that a person British born – and for that matter a New Zealander too – would be honest for less money than one born in Argentina or Southern Europe.

The fact that, for a start, we knew nothing of the language was in some ways an advantage, because we would not be so open to corruption and would be unlikely to fraternise with the local traders. The usual job was called assistant manager, *Ajudante Major Domo*. However, before we could rise to these heights which in those days paid about £1.00 per week and keep, it was essential to learn some Spanish.

I was advised that it was best to go to a non-British estancia and live and work with the peons until I became fluent in the language. I was engaged by the town office of the Mihanovich Co., to go to their estancia about 300 miles south of Buenos Aires; at first as an apprentice but with the possibility of further advancement.

I travelled down by the night train and discovered that although Argentina is a warm country, the nights could be very cold in an unheated carriage and I had forgotten to bring my greatcoat. At Olavarria the next morning I stepped out into the sun and was approached by a youth who said something about *'para l'estancia'* and pointed to a spring-cart with four horses. We drove round the dusty streets – only the central square was surfaced – doing various commissions, and finished up at a big house in a quiet back street.

Here the driver gave me the reins and dived inside. I waited for about half an hour watching the scene. Every now and then a man rode up to the door, tied his horse up and went inside. Just inside the door was a policeman who gave him a cursory search, took his pistol or knife and put it on a shelf behind him. If anyone came out, he was handed back his weapons and he rode or strolled away. Eventually my own driver came out with two others who climbed into the back, took the reins and started off.

This was, of course, the local *kilombo*, or brothel, where for three or four pesos one could have a girl for half an hour. Any thoughts I may have had of making use of this facility were put firmly aside once I started to keep the books and issue out the medical stores. These included permanganate of potash, which was the only cure known for V.D. in those days. On one occasion one of our men had to be sent to hospital with suspected anthrax; on the way he and the driver popped in to see the girls and there the invalid collapsed and almost died. This resulted in the closure of the establishment until it was infection-free, and a claim for damages against the Mihanovich Co., which was still going on when I left.

At the estancia I found the manager, a Swiss who spoke good English, and an Austrian blacksmith who spoke no Spanish. The manager had some rooms on the ground floor of the big house, and I lived in a small hut in the grounds. This had been the original estancia house and comprised three rooms and a cellar. The manager lived by himself; I fed with the men, worked with them, and generally came to understand them.

The horsemen who worked the stock were Argentines of Spanish or Spanish Indian descent. They were the gauchos of popular fiction, and considered themselves a good cut above the manual workers who did the fencing, repaired windmills, made roads and so on. These last were Italian, Spanish or central European and the foreman of this section was an almost pure Negro, one of the nicest men I have ever met. He effected a permanent cure for my racial prejudice. I spent some weeks working under him, and he taught me Spanish in a month.

The food was pretty rough compared with what I had had in New Zealand. Each man was allowed two kilos of meat a day, mainly beef, and precious little else. Breakfast was milk mixed with *maté*, a holly-leaf tea, and some *galleta*, or bread. This was a hard little loaf which used to be delivered to the estancia in bulk every two months. Towards the end of the time it needed an axe to break it and a good soaking in the maté to get it down. Midday was *puchero*, great lumps of beef boiled up with pumpkin.

In the evening there was a roast, ribs of beef on an open fire. Sometimes we had the juice the puchero was cooked in as soup.

The hours were long, from daylight to dark, with a couple of hours off in the middle of the day in winter and up to four hours in summer when it was really hot. After the meals and at any other time during the day we took maté, which was packed into gourds steeped with boiling water and sucked though a tube called a *bombilla*. I have seen it suggested that maté is a drug, in fact it is an antiscorbutic made from the dried leaves of a type of holly, and drunk all over that part of South America. An obvious necessity with that type of diet!

I did find that under this regime I developed an inordinate craving for sweet things. If I went into town, I bought a bag of iced cakes and devoured them straight away. I also bought sugar and put it in my maté. Otherwise, uncomfortable as it was, I was very fit; much fitter than I was when I went to work on an English estancia and ate British-type food with all the trimmings.

After some weeks of labouring I was given a team of about six riding horses and told to understudy the *capataz*, or foreman, when I wasn't needed in the office. This was the real thing at last, and I was fortunate to be on a traditional estancia and so could experience – as few have been able to since – the life of the gauchos as it used to be.

To begin with there had been no ploughing or land improvement. The grassland, *el campo* as they called it, was in pampas grass, huge tussocks with natural grasses in between. It was divided into paddocks of about 1,000 acres and each was stocked with a set number of cows and sheep. Total stocking of the estancia was 5,000 cattle and 20,000 sheep. Each of the gauchos had about three paddocks and he was responsible for the stock. There was very little to do in the way of husbandry; a little hay was made on some improved paddocks near the estancia and this was fed to bulls, rams and pedigree horses. Otherwise the paddock was supposed to keep the same stock winter and summer.

My job was to check the stock and assist when there was a round-up. Each man had a set number of cattle and sheep and had to deliver these to the round-up exactly, or produce the skins or ears of any that had died. I kept a book of what there was in each paddock and tried to reconcile these figures with what we actually counted. Counting the cattle was not too bad. They are big and the mobs were seldom more than 200 or 300. But the sheep were numbered in thousands. For several weeks we dipped about 3,000 every day before breakfast; I had to count these out and try and get them to match the books and the skins each man had produced. By 8 o'clock, it was very hot, very dry and very dusty.

Dipping was against sheep scab, a disease which is now well under control with the latest dips. We had to dip every sheep three times over about nine weeks in order to get any sort of control. Even so there were plenty which had a relapse and these had to be cured by hand which meant catching them in the paddock and quite literally scratching the scabs with disinfectant. Scab used to affect cattle as well: in this case the beast had to be caught with a lasso, thrown and treated.

I did learn to use a lasso but with nothing like the skill of the peons who could catch a cow by the neck or by a leg at will. Once you had the hang of it it was quite easy to catch a cow by the neck. The horses knew which animal you were after when you started to single it out from the mob and followed it without any help from the rider. Keeping the lasso open demands a wrist action much like off-break bowling. Catching the animal is mainly a business of coming up close to the cow or calf so that the momentum of the horse will carry the lasso out to drop around her neck. The lasso is tied to a ring on the horse's girth on the Argentine saddle and, once secured, the cow's neck is nothing like as strong as the horse's middle.

The natural vegetation of the Argentine held a host of wild life. There were rheas which look like small ostriches; at least they were smaller than the African ostrich, although a big one could look me in the eyes. On one occasion I was looking round the edges of one of the shallow pools or lagoons of which there are many, for gull's eggs to supplement my diet when a cock rhea which I had seen brooding his chicks some way ahead of me, attacked. I immediately remembered that I had heard an angry rhea could break a man's leg with his kick, and he kicked me.

He didn't actually break my arm, but he certainly scratched it badly before I had twisted his neck and forced him to stop. He came at me several times before both he and I had had enough. These rheas share the rearing of their young. The female lays the eggs and then the male incubates them and broods the chicks. The male decides just how many eggs he will sit on; so once he has stopped the hen from laying in the nest, she drops the remainder of her output all over the place. These single eggs are called *guachos*, or orphans, and make a good meal or two as long as they are fresh.

The peons were not very fond of eggs, but they were extremely keen on armadillos. These usually lived close to the dumps where we used to put the carcases of dead animals and I was told frequented cemeteries as well. Whatever its diet, armadillo roasted in its shell is very good indeed.

There were also a number of biggish rodents, and we once shot what the men called a *tigre*, a small puma or wild cat. There were wild fowl of every kind around the ponds and a sort of quail or partridge. The whole scene was surveyed by a big vulture-like hawk called a *chimango*.

The plain in these parts is absolutely flat, and the trees were only planted after the Indians had been pacified and the country settled. The dominant features of the landscape were the tall windmills on steel skeletons at the corner of each four paddocks and used to pump up water from deep wells. These, and the odd clump of trees round a homestead or shepherd's hut, were all that one could see in every direction. I remember once some huge wagons came to collect the wool. By climbing a water pump mill by the homestead I could still see them on the horizon three days after they had left with their loads, perhaps 25 miles away.

My time here and elsewhere in the country gave me a great love of the real open spaces. Since then I have seen most of the great plains of the world: the Ukraine, Hungary, the Western United States and elsewhere, and they have never ceased to thrill me as they did even then. There is nothing dull or boring about the plain. The sky at whatever time of day or night is like an enormous dome and is the most striking feature. Riding, driving or even in these days flying across these massive dry seas gives me the sensation of sailing without the misery of seasickness or risk of shipwreck.

Some of my work was to check the stock in the paddocks and I spent many days on horseback watching the hawks to see if they were circling over a corpse hidden in the thistles. These thistles were enormous, and there were great clumps of them through which it was difficult to force a horse. In the old days, before the country was pacified, the highwaymen used to wait until the thistles had grown enough to hide them before they started their operations. I was once riding a grey horse through them and it came out flecked with blood.

If I found a dead beast or sheep I was supposed to take

off the skin, or, if it was too far gone, bring home its ears. As they were all earmarked it was possible to keep the tally of the stock. I once came upon a very recently dead bull, and rather than leave it to decompose and then take its ears, I decided to skin it. I got the skin off in the end and, after a great effort, lifted it on to my horse and then climbed on myself for the 10 mile ride home. As soon as I was in the saddle, he bucked me off and nothing I could do would make him take both the skin and me. In the end I tied it to the lasso and dragged it to the gateway where I was sure I would find it again. Skinning a bull with a blunt knife in the hot sun is no picnic, I got home looking as though I had been in a knife fight, covered in blood, and I felt like death into the bargain.

The peons – they never called themselves gauchos – were wonderful horsemen, riding entirely by balance flat out over the tussocks and armadillo holes when chasing the cattle. I never saw one come to any harm, although horses often fell. They had a trick called *salir parado*, which roughly means to fall while remaining upright. As the horse stumbled or began to fall, the rider would take his feet out of the tiny stirrups they used, lean back and float forward off the horse, ending up running in front or to the side when the animal crashed. I learnt that the only way to keep alive as the horse fell, was to abandon it at the first sign of danger, and I usually had my feet out of the stirrups when working cattle. My horse sometimes came down, but although I was never hurt, I seldom seemed to land on my feet as they did.

The cattle work was fairly rough, lassoing and throwing big calves and horses can lame the beasts. But I realised that the men liked the occasional casualty, and were particularly fond of eating a calf or a colt. If one showed any signs of distress its throat was cut immediately, even if it was only winded. It was skinned, cut up and roasted there and then. A young colt is particularly good eating.

Part of my job was supervising the shipment of cattle from the railway station about 25 miles away. Once we took a herd of cows and big calves spending all day getting to within about five miles of the station. We camped in a side road for the night and cooked some meat on fires made of thistles and cow dung.

We had to drive them through the town to get to the station the next morning, and it took all day to sort the cows from the calves and then load the latter on the trucks. We saw the train off, then started the cows for home. They knew the way back, and the man in the lead had quite a job stopping them from running themselves to exhaustion looking for their calves. I eventually got left behind, and when cantering on to catch them up I fell asleep. It is a most extraordinary sensation falling asleep on a moving horse. I felt I was floating on a sea of daisies – why I can't think. Anyway I was brought back to earth suddenly by the horse putting its forelegs into the skeleton of an animal on the side of the road. He must have been asleep as well.

I once was charged with taking 500 steers to another estancia owned by the company about 400 miles to the west. At first it was decided that they would go by road; then the railway came up with a very low quote promising to do the job in three days instead of the month that the road would have taken. We loaded the cattle on a special train, I climbed into the guard's van and the train trundled south to Bahia Blanca and then to the northwest. The scenery from the van was much the same all the time except for a mountain feature, the Sierra de la Ventana (mountain of the window) which rises vertically from the plain to about 3,000 feet. Its distinguishing mark is a hole right through one of the peaks. I saw this for about three days as the line first went to the east and then to the west of it.

The cattle were to be unloaded and fed at a place called

Catrilo, after two days. I had been warned that the water here was saline and that they should be made to eat hay first. But they were crazy for water and nothing the railway staff and I could do would control them. Some drank too much and collapsed, and altogether half a dozen had to be left behind.

So I did a quick deal with the railway foreman. If he would kill and skin them – they were practically dead anyway – he and his men could have the meat. They jumped at this and while they were at work I went over to the restaurant by the station for a meal. The name over the front was Weatherall and I asked the owner if he was Mr Weatherall. He was indeed, but he had not a word of English. His ancestor had been a soldier with General Whitlock who had captured Buenos Aires in 1806 and then sailed away leaving his garrison to be captured. Most of the prisoners were turned loose in the country and settled down, but he was happy to keep his name pure. He showed me with great pride his great-great-grandfather's belt and sword bayonet hanging on the wall.

We reached our destination nearly two days later, by which time many of the cattle were in poor shape and another dozen were found dead when they were unloaded. As soon as I had got the cattle out, I handed over to the foreman from the other estancia, stretched out on the platform and slept until the manager came to pick me up later.

I spent two or three days on this estancia. It was very different from San Antonio with natural woodland and a good deal of share-cropping. The manager asked me if I would like to have stayed there and manage the sharecropping side, and he promised to ask head office for my services. I agreed to do this if they would agree, but by the time I got back to the home estancia the whole climate had changed.

With the arbitrary ignorance and suspicion that has always characterised the rich, the owners had suddenly decided that they were being robbed. The manager, foreman, even the groom, had been sacked and replaced with a new team. I was kept on as I knew from my book-keeping how the place was organised, but I too was under suspicion. This was because I had been friendly with the manager and had visited with him a small farm in which he had an interest about 100 miles away. Nothing would persuade these people that, as far as the cattle on the place were concerned, they were all accounted for, and that none had been moved. Guilt by association was enough and I was told to go as soon as a successor could be found. The prime mover in all this was the old senora who insisted on my departure in spite of the pleas of her son who wanted to keep me on.

My next job was with a British company. I was now a fully fledged assistant manager or *Adjudante Major Domo*, still on the princely salary of about £1 a week and my keep. But now I had a proper house and a servant to keep it, and fed with the manager and his family in real European style. There were nine estancias in the company: each had a manager and an assistant and were all British except for two who were *Portenos*, or country born. They were barely tolerated in the company of the expatriates.

I got the job because my predecessor had been thrown by a dangerous horse and had very sensibly refused to get back on it again. This was supposed to be a sign of cowardice and he left in disgrace. Soon after I arrived. I was riding with the manager, Elder, when my horse was upset by his Great Dane which killed a skunk under its feet. I was not attending at the time, so the horse got its head down, and started to buck. I had stuck on bucking horses previously, but this beast was a champion and I decided the only thing to do was to abandon ship at once. I came off in one piece but lost the reins and watched the horse, a light chestnut, go galloping off. He must have been a bad one, because he never stopped, hit a fence and hurt himself so badly that he couldn't be ridden. So I shall never know if I would have had the courage to get on again.

These British estancias were very different from my previous place. They had been broken in about 20 years before from the natural campo and sown to lucerne or alfalfa. The new grasses had had a job to survive, and had to be ploughed up and reseeded every few years. For this work we had a team of about 15 ploughmen, each with 15 horses. The practice was that the ploughman used one five on the morning shift, another five in the afternoon shift, and the rest on the following morning.

We had our own staff of blacksmiths and mechanics and even baked our own bricks for the buildings. The manager had built all this up since 1910 and, at the time I was there, the place was running really well. The main income came from the sale of steers, which were all pure shorthorns, mostly roan. They were sold for export to Britain, through a British company of course, by the train load. They were beautiful cattle. The company had its own beef stud which was replenished by purchases from Scotland. At El Dia, I was fortunate to have seen Argentine beef production at its very best.

The trouble was that although the technical side of the farming was firstclass, the economics had been badly undermined by the onset of the world recession. There was no way in which the steers we produced could be made to show a profit at the price we were receiving then; which was, according to notes I made at the time, the equivalent of £9 to £10 per head. In this situation these estancias were very much worse off than my earlier post, as production depended on a constant succession of farming operations, all of which cost money in spite of wages being very low. Only estancias which did nothing but take the natural increase of surplus stock could hold their own. They were only marking time, but they could at least survive.

My boss and the other managers just failed to understand what had hit them. They had evolved this high cost system and now that it no longer worked they just did not know what to do. To make matters worse they were largely on a commission basis – a low wage and 10 per cent on profits. These had been good in the past, but they had mostly been spent and while they could still eat, as the estancia provided all their food, they did not have much spending money.

However they were assisted, if that is the word, by the new overall management of the group which had recently been appointed. These new brooms had been pained to discover that their first year's operations would show a loss, so they hit on the idea of including the replacement heifers, at full market value, in the valuation of stock. Previously stock valuations had only included the breeding cows which had already calved. This move gave everyone a profit – both company and managers – so they got their 10 per cent. Because they got their 10 per cent, they did not draw the owners' attention to what had been happening. The catch, of course, was that the next year the young cattle would have to be valued again, and unless there had been an increase in numbers these would show no increased profit or even no profit at all.

A further example of how big business men work in a crisis happened the following year. Among the estancias was one which had a most unpopular *Porteno* manager who ran it in the old style, by doing nothing but taking the natural increase. As the depression bit more deeply, the management sacked this man – although his was the only estancia making a profit, albeit a small one – and divided his estancia between the two neighbouring ones, one of which was where I worked.

The depression hit others in the country as well. Every now and then we had to give feed and meat to completely destitute farmers who were moving from the south-west of the country where they had settled to areas further north where abundant rain would at least allow them to fend for themselves. This was the same tragedy as was happening in the United States at the time; these families were fleeing from the dustbowl which had been formed by their inability to farm properly, because prices were too low for the crops they grew. These people, many of them were Russian, were travelling with their teams and wagons to an Eldorado which the Government had promised them in the subtropical north. It was customary to feed these travellers, or indeed any travellers, although the management company always questioned it if any of it showed in the accounts.

Nearer at hand there was considerable unemployment, and hardship in the country towns and villages. In consequence there was a lot of cattle stealing; not in mobs, but in ones and twos, which were slaughtered in the fields and the meat or the skins removed. On one occasion we lost a whole field of hay of about 100 acres during a weekend. All there was to be seen of it was a heap in every back yard in the township. But neither the population nor the police were keen to help a foreign company.

The estancia had a small flock of about 600 ewes, and these never showed an increase at all over the years, although no more than about 100 lambs were shown as being slaughtered every year. The Company inspector asked me what happened to the rest and I had to say that the balance were probably stolen by men who would walk some miles on to the place to do this. He suggested that I should set out to catch some of the thieves, but this I refused to do. I knew that everyone was armed, and I had no intention of going out and either getting killed myself, or worse still killing some poor devil whose only hope of feeding a family in those days would be to steal from us. There was even more depression among the grain farmers, many of whom were farming on shares. Wheat and maize would hardly pay the cost of transport to the ports for export, and some trains were being fuelled by grain.

The estancia managers, many of whom had built up their farms to a high peak of efficiency and who had made some savings in the good years, thought that the depression would soon pass, as had others, and that they would be in comfortable circumstances again before long. But meanwhile there was a complete cessation of development work. In particular the practice of renewing the lucerne and rotationally growing grain had been abandoned in favour of cash-cropping every year, or simply leaving the grass as it was. This is the universal recipe for reducing farm output and in the Argentine it was no exception. Production of both grain and cattle which had been steadily rising until about 1930, began to slip. Since then it has never reached the pre-1930 levels. Nor has any fertiliser ever been used in a country which has the potential to out-yield almost any other in the world.

In normal circumstances, reduced output brings higher prices which, in turn, stimulates production and so the cycle continues. But in Argentina politicians of all shades have used the output of the farming industry as a major source of income. The result has been that farmers and landowners have been starved of funds and so never developed the full potential of their holdings.

But all of this was well in the future. While the managers were pretty complacent, I and others of the younger assistants were becoming increasingly concerned. I was 2l at the time, and although Mr Elder – my particular boss – gave me a good deal of scope, it was generally made clear that I should have to wait until I was at least 35 or 40 before I could hope to get a manager's job. I was certain that after a year I could make a good job of managing the estancia, and I could see no sense in waiting until old age, for that is what 40 appeared to me, before I could come into the money.

I was perhaps unappreciative of the good life. There were parties most weekends: for golf – there were several courses on the estancias – for races, tennis and almost everything else one could think of. There were plenty of grooms and servants and there was always lots and lots to eat. The barbecue or *asado* as it is called is an established part of Argentine culture. While the hours were long the work was not unduly demanding, and in my position I did not have to do any of the labouring work which I had had to do.

This pleasant managerial life was in fact doomed. A year or so later the management company got rid of most of the established managers and replaced them with either outsiders or their assistants who came into the job much younger than they thought they ever would. But it was far from being a secure environment for a rather interesting reason.

In Argentina, as in many Catholic countries, the Code Napoleon law on inheritance is rigidly enforced. This means that all the children of a father share equally in the division of his estate. In Argentina, because of the much greater size of the holdings at the time of settlement, the division had not got far at the time I was there. But in the last forty years it has brought about great changes. The Drabble Company used to have three owners for its 100,000 acres. When I was there in 1972, there were over 60 owners. Most of the big estancias have been cut to ribbons with very few holdings of any size left as various inheritors sold to get the cash. This is far from being a bad thing for the economy of the country. Smaller farms mean more intensive production. The total estate, instead of selling off a few thousand fat steers a year as it had in my day, produces grain, seeds and milk for sale as well as beef. This brings work and prosperity to the towns.

I was lucky I suppose to have seen something of the old life of the pampas, though I can't say that I enjoyed it as much as I should have done. I always felt, unlike in New Zealand, that I was a stranger in a foreign land and that the future was insecure. I had no capital to speak of and the chances of getting enough to run a reasonable sized estancia were remote in the extreme. Security of employment, as was soon to be proved, was poor and the politics of the country were beginning the run up to Peronism. The British had always been remote from the local politics, being content to run the trains, the mills, the farms and almost everything else. But it was becoming obvious that this could not go on much longer. The Argentinians resented being second-class citizens and, as a junior member of the British establishment, I heard a great deal of grumbling.

I had two further incentives to move on, although I did not wish to fall back on my parents for any help. Some months before I had received a copy of *Farmer's Glory* by A. G. Street: still I think the best account of farming in the southern counties during the interwar depression. In it the author had shown how he had nearly succumbed and then succeeded in re-establishing himself by new methods. In my arrogance I thought I should be able to do better than he could.

Then Bob Steele, my old boss in New Zealand, wrote and asked if I would consider going back and taking up a share in his property. This looked a fair offer, I was to get a share in exchange for working the place and in some respects it was attractive. But I thought then, and do now, that the place to make a success of farming is one closest to the market. Also I felt that the English were easier to exploit than the New Zealanders.

I was also far from being fit. I had picked up a recurrent fever, somewhat akin to malaria and which indeed was eventually diagnosed as such, together with dysentery. The two of them took me nearly 25 years to get rid of. In the end it was decided that I was affected by actinomycosis, a disease of cattle called wooden tongue in Britain, which is picked up by farmers in the main. It is a long-term disease and very debilitating. It was worrying at the time as I could get no insurance without very heavy penalties because I could never pass a medical. Long after the war a course of penicillin injections lasting a year cleared it up. I never felt well until I was fifty but have never felt ill since. This illness meant that I could never really rely on being able to do a full day's manual work any more, as I certainly had been able to in New Zealand. This meant that I had to rely on whatever other abilities I had to get on.

7. Back in England

Living in the depths of the Argentine I had not realised the full effects of the depression which had really begun to bite by 1932. But on my passage back to London I soon found out. The ship was the *Harpathian*, of about 8,000 tons on which I had a passage for 10s (50p) a day, all found, and to feed with the officers. I shared a cabin with another refugee from the Argentine. The crew, with the exception of three Greek firemen, were all certificated officers or engineers unable to get a ship. The wireless operator, forbidden to send a message for economy reasons, told me that the only communication he received on the voyage, was one cutting his monthly pay from £8 10s to £8.00 and he was married with three children.

Most of these men found their skills were needed in the war a few years later, and paid dearly for them in life and injury. But they were pretty depressed at the time, with no assurance that they would get another ship when they were signed off after this trip. We played poker for matches - no one had any money – from Buenos Aires to Rotterdam, and there was plenty of acrimony over this. The crew ran a bridge school and thought themselves to be definitely superior to the officers, which they probably were in education.

There was no refrigeration, and once the ice box ran out we were on salt pork and beef, ship's biscuits and dried vegetables. This meant that we were made to drink Board of Trade lime juice every day to ward off scurvy. This entailed lining up on deck under the eye of the First Mate. It was completely unsweetened, and I still get a dry sensation of the mouth when I think of it. Having once had scurvy, I did not complain.

On reaching home in 1932, I found the slump had really taken root. There were about 3,000,000 unemployed and farming was in the depths of the inter-war Depression. However, the news from New Zealand was that things there were even worse, with the prices still just about half what they were in England at the time.

Nevertheless the farming depression needs some qualification. Those farmers who suffered most were in the arable districts of eastern and southern England. Again they were, in the main, those who had gone into farming since the war and who had seen their capital quartered in the last decade. If they had been using their own money they were not too badly off, except that they were practically insolvent. Those unfortunates who had borrowed to get established were in a very bad way.

For example, a man who had purchased a herd of cows in 1920 for £80 a head would find that they were worth no more than £20 a head by 1931. Almost every other farming asset had depreciated in the same way. The reason why the arable farmers suffered most was that grain prices collapsed almost completely and the cycle of arable farming is, unlike dairying, a long one, lasting at least a year or 18 months. Even when milk prices were bad they brought in a cheque every month as long as the cows were milked. The arable farms which survived the Depression were those which had the facilities for dairying, a hen unit or pigs.

I should be more specific about this. The arable farms which really suffered were those on poor land away from the sugar-beet factories, which had been set up in the 1920s, and which were unsuitable for growing potatoes and other higher value crops. The derelict land to be seen in many parts of the south was the result of these conditions and much of it remained untouched until the war broke out in 1939. But little of it was really good land of natural fertility. Even in the depths of the slump, it was not easy to buy what one would call a good farm; the owners or tenants of these managed to carry on somehow.

My own prospects at the time were not particularly good. I still had the £500 I had started out with and about £100 besides which I had managed to save in the last four years by the exercise of incredible meanness. My father offered to lend me £3,000 more when I decided I should make a start, and advised me to get some more experience while I was looking round.

I certainly needed the experience. My skills of shepherding and looking after cattle on the New Zealand and Argentine pattern were hardly suited to the small farms and careful husbandry of England. The one essential thing I had learnt in my travels, which stood me in good stead in later years, was that success in farming lies in knowing not so much what to do to get the right results, but rather how little one could do and still get away with it.

In many ways this is a negative approach. But in hard times it is often the only way to safety. The trouble is that, once learnt, it is impossible to forget, and I have drilled the need for this attitude into my sons so well that one of them accused me the other day of having brought him up to be too mean.

I had been impressed, perhaps overimpressed, by *Farmer's Glory*, in which the author had succeeded in overcoming the slump by milking his cows in an open-air shed on the downlands of Wiltshire. The system he used had been invented by Arthur Hosier who farmed on the Downs above the Pewsey Vale. Its method was as follows. Instead of having good cows in rich meadows tied by the neck in stalls for most of the winter and milked by hand, Hosier's way was to milk large numbers of very cheap Irish heifers on poor downland, feeding them with grass, hay and cheap imported feeding stuffs.

After a few years of this sort of farming, the pastures improved from the manure and treading and the consequent increase in fertility allowed for heavier and heavier stocking.

Hosier found that, by this means, he could cheapen his costs and thrive at a time when many dairy farmers on better land were having a hard time. The Irish heifers, mainly from the poor hills of County Cork, improved out of all knowledge. After two or three years they grew much bigger, and it became the usual practice to sell them after three calves to be kept on the more traditional dairy lands. This system avoided the perils of depreciation to some extent; herds were always kept young, and the farmers were not forced to rear their own replacements.

Hosier, who was a most practical inventive genius, produced many other implements and ideas to foster dairying in the Depression and established a factory on his own farm for making them. Many of his outdoor bails, as they were called, were used on suitable land and today there are still some to be seen in Wiltshire.

However the system was still looked down on by the traditional dairy farmers, especially those on heavy wet land where it would have been impossible to winter the cattle out of doors. This was not because they would have come to any harm in bad weather, all farm animals are well insulated, but because the pastures would have been destroyed by their hooves.

My old mentor, P. G. Holder, had asked me to go and see him on my return. I found that he had now directed his enthusiasm to large-scale dairying and was planning to erect a cowshed for 288 cows: would I go into partnership with him. My parents thought this a good idea, just the thing for a young man to be associated with a rich and established one. I didn't see it that way at all. Unless, I told him, he would be prepared to give me control of the set-up, I would sooner go off on my own.

This wasn't at all to his liking. The pleasure he obtained from his various schemes was from their planning and initiation: he just hadn't a clue or any interest in running a business methodically. He knew at once that his enthusiasms, which I was sure would be recurrent, would be effectively blocked by me if I had the sort of control I wanted. We decided it would not work, but there and then he offered to back me in any venture I wanted to undertake. I never took him up on this, but it was a great boost to my morale.

I did agree, however, to help him set up the new dairy herd and spent nine months learning all about cows in the mass at his expense. The cowshed at Peaton Hall near Ludlow was one of the wonders of British farming in those days. It was built to the latest patterns by a number of contractors and suppliers whom Holder had managed to bulldoze into giving him very favourable terms indeed.

There were certainly problems. He was buying the cows from Ireland, and every week there was a message from the station at Craven Arms saying there was a truck of cows to be cleared. These had to be walked home and usually one or two had calved on the way. The building however was not progressing as fast as the cows were coming in and so all sorts of compromises had to be evolved to get them milked in relays in the few finished stalls. None of the men – and there were a great many – were skilled dairymen. They could milk, but had no idea of how to use the machines or keep them clean. It was a continuing nightmare, or challenge, depending how you looked at it. Every time some sort of system was established, P.G. would produce another 50 cows or the plans for yet another larger cowshed and start ordering the equipment for it. The eventual total came to well over 400 cows.

However by the late spring I was saved from too deep

an involvement by a recurrence of malaria and by my insistence that, as I was looking for a farm of my own, he really should get someone else to look after the cows. He immediately offered me the tenancy of one of the farms he owned, quite a good farm on heavy land. The main difficulty was that it was within a few miles of Peaton Hall and I knew full well that I would be pressed into managing the cows whenever, as frequently happened, he sacked his head man. He would never sack me.

I was also obsessed with the Hosier system and the attractions of cheap land on the Wiltshire Downs and had spent any spare time that I had looking for farms to rent there. On reflection, I was probably mistaken in this. I have always been keen to take on land by price and not by its intrinsic quality: the result has been that I have never farmed a decent bit of dirt in 50 years. I could have taken a farm with good land when I started for very little more money than the bad or second-rate.

Finding a farm to rent was not too easy. My manner with the landlords or their agents was far from being ideal. I knew little about English farming, had very little money, and I was rather young (23) trying to get into a farm when many older and more experienced farmers' sons were trying to get into banks and other safe city jobs. Most landlords, very reasonably, would much sooner let their farms to established farmers or their sons, than to complete and presumptuous strangers. They are still of the same opinion today. I motored thousands of miles looking for a farm and generally found that they were becoming vacant because the farmers had found the going tough. The buildings were falling down, the fences were useless and the arable land, if there was any, was full of weeds. In many cases, the crops I saw were hardly worth harvesting.

Still, people were taking land. Hosier accumulated thousands of acres in this period, as did a number of others. It did seem that, under his system, all that was needed was dry land, water, and strands of barbed wire to contain the cows; and eventually you could build up a farm. I used to tell Holder that I could rent land in Wiltshire for a quarter of the reduced rent he wanted to ask me. He used to laugh and said that if I settled in those benighted parts he would never come and see me. He never did.

Eventually I happened on Manor Farm, West Knoyle—about 500 acres which had failed to find a buyer at auction. It was owned by one of the successful Wiltshire farmers and had been occupied by his brother who had given up. I approached Tom Coles, the owner, and offered to rent it. The rent was fixed at £320 a year; about 13s (65p) an acre, which was considered a high rent for those days. It had a seven bedroomed house with few conveniences, four good cottages and the usual buildings. The land was about two-thirds light chalk and the rest heavy clay. Being right on the western edge of Salisbury Plain, it had a steep road right up through the farm. This hill, I was told, was the farm's worst drawback because at least three horses were needed to pull a ton up it.

My taking this farm had caused quite a stir locally I discovered later. Someone else was trying to take it for his son, but he had been so devious in his machinations that when I had come on the scene and not argued about the rent the owner had closed at once. I was warned though, as soon as the news got around—which was about ten minutes after the deal had been effected—that it was the worst farm in the district, that no one had ever succeeded there, and that I was doomed to failure.

There was nearly a failure of another sort. I was at the time engaged to be married and, from time to time, I had taken Ronie around with me when looking for a farm. On this occasion she hadn't been with me on my first inspection but came with both our mothers when I clinched the deal. I negotiated in the car with the landlord while Ronie and her mentors looked round the house. When they came out I asked them what they thought of it. I had seen no more than the kitchen myself.

'Impossible,' said both mothers, 'it won't do at all.'

'In that case,' I said, 'it is just too bad. I have just taken the farm. . . .'

8. Manor Farm, Knoyle

I moved into Manor Farm, West Knoyle, on 29 September 1933, with no very clear idea as to how I was going to run it. I now knew how to manage cows indoors, but had no intention of doing the same. I also knew something of the arable work which would be necessary because about 150 acres were under the plough. In the event this was not to prove very difficult. In those days the arable sector was in a four-course rotation of wheat, fallow or roots, barley or oats, and clover. My predecessor had left the farm in its rotation and all I had to do was plant the crops needed. My tenancy agreement laid down that I must stick to this rotation, as did that of every tenant farmer. Farms could be lost for not keeping to it. As the Depression wore on and farms became harder to let, landlords no longer insisted on these rights but they still held good in those days.

There were seven men on the farm when I took it and I decided to keep three and look for a cowman elsewhere. The three were a young carter, his father and brother. At that time the weekly wage was 28/6d (£1.42½p) for a 56-hour week; the men had no set holiday, but could plant their potatoes on Good Friday. The carter was a brightish young man and worked for me for many years as a tractor driver.

My predecessor held his sale and I was able to buy most of the implements needed to run the farm for less than £100. This also included the best of the carthorses, a gelding called Duke which made nearly £30. I bought a few of the heifers there for between £12 and £15 with

Capital available September 1933.
  My own   £500
  Father's loan up to (bank guarantee)   £3,000
                                           ──────
Total   £3,500

	£	s	d
Expenditure going into Knoyle			
Valuation	523	5	8
60 heifers @ £18	1,080	0	0
20 ,, @ £15	300	0	0
Tractor	100	0	0
Bail	180	0	0
Plough	30	0	0
Furniture	180	0	0
Pigs	70	0	0
Sales purchases	100	0	0
	2,563	5	8

	£	s	d
Balance for seed, feed, working capital	936	14	4

In 1935 left Knoyle with:
	£	s	d
Valuation	623	0	0
Paid Peterson (Valuation)	157	0	0
75 cows, ex Knoyle			
60 heifers for Chute			
50 heifers			
Implements etc. for 900 acres			

calves at foot and had to milk them myself in the old cowshed as I had not yet bought my outdoor bail. Nor had I got a cowman.

Greatly daring I bought the first tractor in the parish from a famous farming pioneer Rax Paterson. He was a disciple of Hosier who had this new Fordson for which he had no work. It cost £100. Rex in many ways surpassed Hosier in his expansionist plans. Starting with 60 cows and a bail on a patch of hill land in the hills near Chute in Wiltshire, he built up to more than 10,000 acres in the south and in Wales. He, like Hosier and many others of us, specialised in adapting run-down arable farms to dairying when, as in the case of Manor Farm, the previous owner had gone under. This cheap and extensive dairying worked well and was profitable in the 1930s. However since the war the rising cost of land, livestock and everything else means that intensification must now be right.

Rex also found me my first cowman, an Australian called Clyde Williams, and I bought a second-hand bail at a farm sale rather than spend the £300 needed for a new one from Hosier. I also picked up a total of 80 calf heifers averaging in price around £17 and was able to start in business.

It was a simple system. I rose at about 5 a.m. and walked up across the farm to the field the cows were in. Somewhere in front of me I would hear the petrol engine of the bail start chugging away and the lights in the milking section came on. If they had not been aroused beforehand, the cows heard the noise of the engine and began to get up and stream towards the bail. The engine was their dinner bell: they knew that, once in the bail, they would get a helping of feed while they were being milked.

Clyde and I drove the cows into the temporary fence which held them and there they stood patiently waiting for their turn to come in and be milked. They soon developed a sort of pecking order, and it was not long before they came in a regular sequence. They were all given names and, as I have always had a good memory for animals, I could name up to 150 cows and know them all apart. Cows have a lot of individuality in their markings and conformation and a good cowman should know them all as individuals and not as numbers. Numbers are handy, of course, if the cowman should be taken sick or be on holiday and are essential to any record-keeping; but there is something very impersonal about identification by number alone.

In those days it was considered essential to strip out the cows by hand before turning them out after the machines had been taken off. This process produced very little milk but it was generally considered to be the cream. The main object was to make sure that the cows had been properly milked out and there was nothing left in the udder to cause mastitis and other illnesses. In this way a very good check could be kept on the state of the cows' udders and their general health. It was also rather pleasant on a cold winter's morning to snuggle up to the cow's flank and listen to her stomach rumbling as she gobbled up the last mouthfuls of feed.

Before I finished keeping cows some 30 years later it was generally held that stripping cows by hand was a waste of time and they could just as easily and more quickly be stripped out by machine. Today it is true to say that few farmers or dairymen need to know how to milk. As milk yields are immeasurably higher than they were then, there cannot be much wrong with the system.

As milking was finishing, one of us would go out to catch old Duke and harness him to the float. Then we loaded up the churns to be brought to the farm for the lorry to collect. During the morning, Clyde would sterilise the machine with steam from a portable boiler, haul out supplies and perhaps move the bail on to fresh ground.

The beauty of the system was that, even in bad weather, it was possible to move on to fresh land at once. There were no acres of concrete to swill down.

Very little fertiliser was used in those days and it was generally held that the manure dropped by the cows and the treading of their feet improved the pastures so that eventually they reached quite a high standard of fertility. This could then be cashed in by cutting a field for hay or carrying more stock. The system really depended on feeding very cheap concentrates, mostly imported, and meant farmers became such experts at this concentrate feeding that they simply used their pastures as exercise grounds for their cattle.

After a while I engaged a boy to help Clyde on the bail. Sometimes disasters happened. These usually concerned the engine. Petrol-paraffin engines were very unreliable in those days, and there were occasions when the engine would not start. When that happened I used to turn the whole staff on to milking while we wrestled with the engine. It was essential to get the milk away, especially in the summer, as otherwise it would go sour and be lost.

Then there were calving problems. In those days it was the mark of a slack farmer to go to bed while a cow was on calving or even looking as though she would. The calving cows would be kept in a paddock by the house and I would look around them last thing at night. Any sign of trouble, real or imagined, meant getting her in and assisting her. I was quite a good amateur vet and spent many nights helping a bad calving case with the rest of the staff. But, like stripping, too much attention to calving went out of fashion and, in general, it can be said that most cows will calve by themselves if they have enough time. The less attention we gave the down calvers the less they seemed to need, which as my wife said was what we should have known in the first place.

My predecessor had not sown any roots although he

had left me a lot of hay to take over by valuation. He had also planted a field of rye, which was supposed to be eaten by sheep to improve fertility, according to the rotational rules under which he and I had to farm. I had no sheep at the time, but I had a good friend and neighbour called Jack Stratton, one of a large and most successful family of Wiltshire farmers.

Uncle Jack was a bachelor, and had been blinded in an accident while still a young man. Although he regained some of his sight, he never really saw what was going on around him. He lived with another brother not very far away, and I often used to have supper with them, paying for it afterwards by playing bridge with the two old gentlemen and their housekeeper. I used to make them very cross. I had been up since 5 a.m. and was dropping off to sleep as soon as I had eaten, but I used to stick it until I could decently go home without revoking. I have never played bridge since.

Anyway I told Uncle Jack of my problem, and he said at once that he would send his sheep over from Pertwood, a farm he had nearby. Eventually the rye was fenced in and a huge flock of sheep arrived, scoffed the lot in about four days, and were driven away again. I saw Uncle Jack at least once a week after that but no mention was ever made of payment for the keep so I broached the subject one evening over supper.

'You have a lot to learn,' he said. 'Don't you realise that far from you doing me a kindness by keeping the sheep for some days, I am owed quite a bit for the fertility they have left on your farm. If it wasn't for our friendship, I should be charging for the manurial value according to the custom of the country!'

I swallowed this with grace and put it down to experience; but I was never able to persuade people to pay me for feeding off their crops. Although blind, Uncle Jack knew how to survive in a bad time. Uncle Jack was one of the great characters of the district and I was involved in his affairs until he finally died at the age of 92. As the affair of the sheep showed, he could be as sharp as any of his neighbours in business. But when it came to personal affairs he was generosity itself, and had assisted a number of deserving and undeserving cases.

At one time he had been a very good horseman and even while blind would go riding at great speed over the most unsuitable country. I used to go with him occasionally, and his great joy was to take off across the downs leaving the horse to find the way. I, who could see where he was going, was petrified but he never came to much harm. His greatest interest was in disasters such as happen on every farm. He usually had a nephew or a manager actually running his farms, and naturally enough they did their best to hide their problems from him. However he had a knack of finding out what was happening, and would remember these long after everyone hoped it had all been forgotten.

I remember I drove him one stormy Saturday right down into Dorset to see some piece of farm equipment he wanted to buy. As we came home he began wondering out loud if his shepherd had remembered to put the sheep, which had just been shorn, into a barn to escape the heavy rain. We duly went back to the farm, the shepherd had not put the sheep into the barn and a number had died. I saw Uncle Jack about a week before he died – which was years later – and he went through every detail of that drive, its aftermath and the number of ewes that had died.

The farming scene at that time, although well publicised as being depressed, was remarkably cheerful. Nearly all my neighbours milked cows and there is no doubt that for several years in the early 1930s the monthly milk cheque kept most farmers afloat. There were still a few arable farmers clinging to their farms, but most of them in Wiltshire anyway had reserves of their own to carry them through. Things did not come good for them until the outbreak of the war, or a year or two before.

I must have been born with an acquisitive or expansionist streak because no sooner was I settled in Manor Farm than I began to let my tractor and driver out for contract work. I also began looking for more land. There was not much money in contract work, but wages being so low, there was always something to keep the cash flow going, although in those days it wasn't called a cash flow. After a few months I took another 80 acres off a neighbouring estate for £20 a year including some quite good buildings. In doing so I outbid everyone else and made them very angry. I kept 150 fattening pigs in the buildings, made hay, and sublet the grass for sheep for about £30.

I also felt that the price for summer milk at that time – about 8d (3½p) a gallon – was too low and decided to make Devonshire cream to sell to the motorists who were on their way to and from the West Country. There was a petrol station on the A303 at the top of the farm and they agreed to sell it for me, except when they were very busy when I came up to do it myself. The principles of making clotted cream are well known. The milk is set in wide dishes and then the cream is helped to rise by raising the temperature. Once it has formed on the surface, it is skimmed off and put into pots. In a busy household there is seldom time for careful attention to detail and the milk sometimes came to the boil, with no ill effects.

I doubt if I made much worthwhile money by these activities, but they kept me occupied and brought me into touch with the commercial side of farming. I liked having a deal on, even if it was only for a brace of rabbits to the game merchant. The principles of trade are the same whatever the size of the bargain. It is becoming fashionable these days to say that farmers should hand over their sales business to Co-ops, groups and other experts in this line.

The farmer's job is to farm, he has been trained for that and so therefore he should stick to that, say some experts.

I cannot agree. Part of the failure, as I see it, of the New Zealanders to exploit their farming really to the full, is that the average farmer there spends the whole of his life on the farm, and leaves all his buying and selling to specialists. He can gain no experience of what business is all about. I used to go to market regularly, something few of my neighbours do today. In fact those that do attend are usually of my generation. I think it makes for a duller life without the companionship and gossip in a market once or twice a week. It can, of course, be turned to profit. You can hear that someone is short of keep and wishes to sell some stock, or someone else has a rick to sell or you can let your wants be discreetly mentioned.

There were some old-fashioned pigsties on the farm, and sufficient empty stalls for about 60 cows. I had always had an interest in pigs, so I bought a dozen gilts for £5 apiece, a boar, for a tenner, and I was in business. I farrowed the sows down in the old pigsties and ran the young pigs, once weaned, on a bracken bank not far from the buildings. During the winter it was easy to see how they were going on, but once the bracken grew in the summer they were often very difficult to find.

They weren't the only inhabitants of the bracken; a family of badgers were established there and when chasing out the pigs to put in the old cowsheds for final fattening, they were sometimes driven into view as well. I don't know how economic these pigs were. A baconer used to sell for about £5 and my feed mixture used to cost about the same per ton. Today a baconer is worth up to £50 but feed is £80 per ton which makes it look as though the economics were probably better then.

Of course prices were abysmal. I was buying wheat delivered at £3 10s per ton with barley about the same. The wheat I grew and sold made about £9 per ton that year as it was subsidised under the Wheat Quota arrangements by a levy on imported wheat. The milk price for my first year averaged 10*d* (4p) per gallon. My cowman's wages were £2 5*s* (£2.25) for a seven-day week; his assistant got £1, out of which he had to pay his board. Today milk is worth 50 pence and a good cowman would earn nearly £75 for the same hours.

For the first few months, until I was married, I lived off the farm. At one time, I stayed with a neighbour, Peter Smithers, who had a farm about five miles away. He was very keen on horses and placed hurdles in the gateways, instead of gates, so that he could jump his way round the farm. I remember riding round with him one afternoon and noticing that he sat very firmly on his horse when he took the jumps and kept his feet right into the stirrups.

He laughed when I told him I thought that taking jumps without a companion who could get help in an emergency could be dangerous, and that this would be compounded by sticking so firmly to the saddle. He would be unlikely to be flung clear, should the horse come down. The very next evening I came in to tea and was told that Peter had tried to jump a stile, had fallen under the horse's legs and was in a very bad way. He died that night.

This was a dreadful shock for me. I had done a lot of riding, and had many falls and seen others, but it was the first time I had lost a friend in this way. It rather turned me off riding, particularly as by then I was finding the ground a great deal harder than my head. It's a funny thing that, until about 25, the human frame can take any amount of punishment without permanent harm but after that falls come hard and they are best avoided.

However, it was, and still is, a horsey district and the farm enjoyed – if that is the word – the possession of a large wood which was a sure refuge for foxes. Hang Wood drew foxes and, as the local hunt was out about three days a week, there were visits from them every few days. There is something particularly intolerant about a hunting man or woman and, in those days, they considered themselves to be the lords of creation. On one occasion Ronie was in the house putting up curtains and, without asking me, the hunt met outside. One elderly lady asked me to hold her horse while she got off. She then went indoors and without saying a word walked right past Ronie upstairs to the lavatory. She duly emerged again, remounted, and rode off without another word. I knew the English upper classes were inarticulate, but this was ridiculous.

Some of the land on Manor Farm is very heavy and the passage of a hundred or so riders after heavy rain did a lot of damage. This infuriated me because there was never a word of apology, or a request for permission to ride on the farm. Certainly allowing the hunt to come in was not in the tenancy. If people had tried to do this in New Zealand, I am sure that they would have been ordered off. So, greatly daring, I wrote to the Master and asked if he could see his way to keep off the heavy land, particularly in the spring, and out of the fields where the cows were running. I say greatly daring because most of my neighbours, who used to complain as much as I did privately, refused to do anything publicly.

The response was rather interesting. I first had a call from a couple of the hunt supporters, rather crusty county types who asked if they could buy some straw. As straw was a drug on the market at that time and I had taken several ricks in the valuation, I was able to oblige. Then the Hunt secretary turned up and asked if I had lost any poultry to foxes, and to let him know as soon as there was any trouble on that score. Another member came to say that, if I would like a day out, a horse would be provided.

As I had been rather ungracious to these emissaries, I was then visited by the Master himself, John Morrison later Lord Margadale and a man of considerable charm and reason. Yes, he saw my point about too many horses too often. He would see to it that as far as it was possible, visits would be limited. In addition, as I had told him of my interest in deerstalking, he offered me the freedom of one of his woods in the culling season.

When I told my neighbours about this, they thought I had been very daring to protest at all. In those days, tenants had been turned off their farms for being bloody-minded about the hunt or the landlord's pheasants. The fact that my landlord's son was the Master's Agent made my action even more foolhardy in their opinion.

However, soon after the hunting season ended, my landlord told me that for family reasons he would have to sell the farm. He had an offer for £5,500 – just over £10 an acre. If I could match that I could stay; if not he would have to give me a year's notice to quit from the following September. This was a blow which I found hard to bear at the time. I had recently married, we were expecting our first child the following November, and I knew I would have eviction hanging over me for the next 15 months.

This bombshell completely altered my farming plans. I knew that, whatever happened, I should have to give up in September 1935 and that I would have to farm to leave the new owner the arable in rotation. The dilapidations for unpainted barns, broken gates and ragged hedges would have to be avoided by doing the maintenance which I had been postponing until I had got established. But the farm had to be farmed every day until I left, and I had to spend some time looking for another place.

The farming went on as best it could. The grass grew naturally in the spring without my spending money on nitrogen so the milk output rose, pigs multiplied and it wasn't long before we started haymaking. This used to be quite a performance. Gone were the days when hay was pitched into the wagons by many men. I was mechanised.

The grass was cut by tractor; the mower was one of the few new implements I bought. The hay was turned and raked up by horse-drawn tackle, and then swept to the elevator with a car sweep.

This was another of Hosier's inventions. The motive force was one of the many high horse-power old cars, which could be purchased for a few pounds and which still had plenty of life in them. The sweep fitted on the front springs was easy to handle, and the hay was simply swept up to the elevator and raised to the rick. The motive force here was an old horse walking round and round driving a spindle. It was quite hard work pitching on to the elevator and then forking it out again on the rick.

I used to help on the rick, and always had a pocket full of pebbles with which to encourage the horse to keep going. I made hay in this way for about the next ten years, putting it into quite big stacks. Sometimes the hay was a bit green and the stack heated. But I never had one catch fire, although on occasions I had to cut them open to let the heat out. Once I even turned one. That is, set the elevator at one end and just rebuilt the stack.

Haymaking problems were a staple subject of gossip on the farms those days, and when in trouble we used to try and hide it as much as possible. But this is difficult, the smell of heating hay is very pungent and carries a long way; I can recognise it anywhere. If we were seen turning a rick, there were always several phone calls asking if I had found my watch yet, the usual excuse. Today nearly all hay is baled in neat little parcels and while it doesn't catch fire, it goes bad very easily. I have never really made good hay in bales. There is something about hay curing in a big rick which seems to improve it.

1934 was a very dry summer, one of the driest I can remember, and hay crops and grass were both short although I had enough to carry my herd through the next winter. Grain crops were good, particularly wheat, and I had the best yield that I had for many subsequent years. A hot dry summer usually means a good wheat crop, and yields as I remember were about 23 hundredweight an acre. One neighbour of mine used some extra nitrogen and secured 2 tons an acre, an unheard-of yield which no one believed except for myself. I had helped him thresh the rick in this particular field.

Because of the dry weather it was an easy farming year, and I was able to take time off to go and find another farm. I was still sold on keeping to the light chalk lands in the south; although Mr Holder, hearing of my predicament, offered to let me one of his. However I was adamant in my insistence on keeping away from his schemes, which I was sure I would become involved in. However good farms were not easy to find. Those available were usually arable farms without water and grass for stock.

Arable farming had such a bad name in those days that only a brave and foolhardy man would have risked going into one. Some of them were to let very cheaply, rent free in fact, and the landlords would advance the money for the valuation. But the earnings from grain were so low that they did not provide enough of a cash flow and without cows or a private income there was not a hope of success.

I was offered and refused several of what are now the best farms in Hampshire, and have often kicked myself for it. But they were without water and, even if water had been available, it would have taken at least a year for me to have got enough pasture going to keep a herd of cows. In those days I could only think of cows.

There were other problems. The purchaser of my farm, who had bought a good deal of land elsewhere in the district, was an Australian or of Australian antecedents. He was a persuasive gentleman and managed to get the solicitors he employed to pay the deposit on my farm and on others he had undertaken to buy. Questions began to be asked about the soundness of his finances. In the end he could not pay for the farm and it was offered to me for £4,500 which for the size looks ridiculous today.

This was in the early summer of 1935. I had £500 in cash at the time and I suggested to the bank that, if I put this down, they should provide another £4,000. The local manager agreed, but the application had to go to head office where my father was in charge of advances. He refused the loan on the grounds that the bank would only lend two-thirds of the purchase price, namely £3,000. He would, he implied, lend me the balance himself.

This was nonsense to my mind. The farm had originally made £5,500, and I had the benefit of the £1,000 deposit that the solicitors had forfeited when their client failed to complete. In any case my pride or obstinacy made me decline the offer and I determined that I would not borrow from him or the family, and would be independent instead.

I nearly had to eat my words. My landlord, to whom I explained my position, was sympathetic and agreed to hold the offer open until the bank came to its senses or rather my father did. I think he would have left the money on if I had not been able to raise it elsewhere. Jack Stratton also said that he would buy the farm if the worst came to the worst. So I soldiered on until, in August 1935, my landlord suddenly died and his family needed the money immediately.

I was still frantically looking for another farm and was given an ultimatum: either buy before the 15th of September or get out on the 29th. My reason for not buying was not simply pride. I knew that Manor Farm was not a good farm, and that there was little scope for expansion at hand. After taking the 80 acres next door, my neighbours were all watching me like hawks and there were some pretty big fish among them. I wanted room, and the chance to expand, and I sensed the opportunity for this was likely to be where cows were little known and farmers were badly off. The dairy farmers of the Vale were solvent and keen competitors. I had seen plenty of derelict farms in Hampshire, including one 3,000 acre block which the owner refused to let or farm in any way, because of tithe. If he did anything, then either his stock or his tenant's would be distrained.

On about 13 September, an acquaintance rang up and asked me if I still wanted a farm. I said I did and arranged to meet him at Tangley the next morning. The situation was interesting. The farm had been unstocked for a year and there was a summer's grass on about 1,000 acres. He had taken 400 acres with a house and the landlord insisted that someone else should take the other 600 acres or he would not let it at all. There were snags. The tenancy was for 11 months only with no real certainty of renewal. But I liked the farm, and I liked the area. There was water, the grass was good, and there was enough hay to take the cows through the winter. The landlord agreed to give me an option for the next year on the same terms and I agreed to take it.

My friend and I repaired to the local pub and were having a drink, when Rex Paterson, whom I knew well by this time and who lived not far away, saw my car and came in to see me. He asked what I was doing and when I told him that I had taken Tangley, he said that he had always wanted to take the farm himself. I didn't tell him the terms of the lease. He then suggested that I should take Greens Farm at Chute on which he had been milking about 70 cows, and which had quite a good house and a couple of cottages. I agreed there and then, and we initialled an agreement on the bonnet of his car.

I then came home and told Ronie that we would be leaving in about a fortnight. With the fortitude with which she has always put up with my actions through our life together, she made no objection and started to prepare for the move. It was one of the best decisions I ever made in my life. It got me into a district where there was no competition for land, at a time when it was at its cheapest. It is not good land, being too stony and too stiff with clay for that, and it is at a high altitude, but it has done me no harm.

9. Greens Farm, Chute

Leaving a farm is always a wrench; in many cases it is a mark of death or failure. In my case the discharge, as they say, was honourable. Under the terms of my tenancy I was liable to a year's notice at any time, and the fact that it came so soon after my entry was just bad luck. I had the same experience a few years later and this made me really determined to do what I could to make tenancies as secure as possible. In this I was fortunately successful and tenants and their sons now have security, although for my part I don't benefit; long ago I managed to purchase all the land I farm. Again this was a departure. In those pre-war days, most farmers had no thought but to get a good landlord who would do a great deal for them and charge a minimal rent. No doubt if I had farmed on such an estate I would have thought differently, but as I never experienced a good landlord in the accepted sense of the term, I felt I just had to buy.

But then I was lucky in my date of birth and the district I chose to live in. The 1930s were the nadir of the inter-war depression, particularly on the southern arable lands. They were not soils where potatoes and sugar beet could be grown, and were of little inherent fertility. Even if they were fenced in and allowed to fall down to grass, they never produced feed to equal the stronger soils.

I moved from Knoyle to a total of 900 acres of grass when I came to Hampshire. For a year or two all the stock I had were 140 cows and about 60 young cattle. I used to buy all the feeding stuffs and even some hay. My friends from Gillingham vale – and Manor Farm used to have some vale land—would keep that number of cattle on less than 250 acres. However, these infertile arable farms provided magnificent opportunities for a host of incomers from other districts and from outside farming altogether.

There was another advantage in coming green to the area. I had no preconceived ideas at all. My farming mind was blank: I knew nothing of fertilisers, seeds and varieties, as my two years at Manor Farm had hardly given me a deep insight into arable techniques. I had just followed my predecessor's rotation without much thought. At Tangley my initial start was milking cows. I had 70 which I had brought up with me from Knoyle together with their bail and the cowman, Stan Nokes. He had joined me the year before, and is still pottering about more than 40 years afterwards.

At Chute, which I had taken from Rex Paterson, there was a fixed as opposed to a moveable bail. I found that, after I had secured my outgoing valuation and sold my pigs because there were no buildings, I had a credit facility of £900 with the bank. I bought 60 Irish heifers for about £14 a piece, paid Rex £100 for the bail and bought compounds and hay as required when the milk cheques came in. In fact I stocked a viable farm for £900 or £3 an acre.

There was a keener atmosphere in Hampshire in those days. In west Wiltshire and Dorset, the cowkeeping farmers were established and complacent in spite of the Depression. Few of them, or their sons, risked venturing the few miles east towards much cheaper land, although there was quite an influx from Devon and the north of England where farms were small and families large. But most of my neighbours in North Hampshire were newcomers and many had no farming roots at all.

We had a great example to follow – Arthur Hosier. Hosier was the high priest of cheapness and improvisation. His milking bail, simply a machine on wheels, transported dairying out of the vales to land where no one ever thought cows could live. He himself was a farmer's son and was sufficiently eccentric to break with tradition and get milk from the hills. He built a tremendous farming empire with cheap land and cows and, like all farming empires to date, it eventually failed to outlive its founder and dynamic force.

He was never selfish with his ideas and would discuss his system with anyone interested. Of course he hoped that the newcomer would buy some of his machinery, but failure to do so never seemed to make him any less expansive when meeting him. Farmers are always tremendous talkers: quite a lot of it is boasting and exaggeration of yields and the performance of their stock. But, unlike what we hear of industrialists, they don't hide their successes beneath the wraps of secrecy. Most of the pioneers, and there were quite a few in those days, would welcome interested visitors and never minded imitation. Hosier, it is true, formed a machinery company to exploit his inventions, but his prices were so reasonable, and his service so good that even the meanest of his customers – and farmers can be pretty mean – never grudged him his profit.

Hosier employed his brother-in-law, Charles Whatley, as the technical expert, who transformed his ideas into reality. If Hosier wanted to develop a gadget to cool the milk, Whatley made it possible. He was one of the best practical mechanics I have ever met, and would turn an old mowing machine into a self-propelled grass-cutter by marrying it to one of the plentiful supplies of old cars which were the backbone of much of our mechanisation. The favourite car for this conversion was the Model-T Ford, but as these were already something of museum pieces by then, others had to be used as well. I made quite a good tractor from a Sunbeam Laudellette and until the clutch gave out after about a year's work, it used to chain-harrow the grass and haul the milk about if the horse happened to be lame, or otherwise indisposed.

It is sad to report that when Hosier reached a fairly advanced age, he sold his manufacturing business without mentioning it to Whatley. The latter then took charge of his books of designs and refused to give them up. Whereupon the two old gentlemen, both in their late seventies, fought for it in front of the staff until they were exhausted.

The Hosier system brought dozens of otherwise hopeless farms through the Depression because it made dairying possible on land where previously nothing viable could be produced. Milk prices were not good, but they provided a regular income every day of the year, and this is a very comforting thing in a farming business. I have never, in my time, known even a moderate dairy farmer go wrong.

My move to Hampshire made a complete break with traditional dairy farming. Although at Manor Farm I had used a milking bail, I was still in an area where old ideas died hard, and I used to graze the cows on the vale land in the summer. Tangley was, and is, reasonably good for growing grass, thanks to its altitude. Most of the land lies over 600 feet above sea level and rainfall is higher than at Winchester 20 miles to the South. Even in dry weather, there are frequent morning dews. I read Cobbett's account of these parts some years ago, and he remarked on the grass meadows of the high hills with approval. I did not know this at the time of course. What had attracted me to the place was that it showed a good stand of grass because it had carried no stock for the previous year, and the owner had made very little hay. The Chute farm was much the same, but had had the advantage of being farmed by Rex Paterson for several years. The pastures, thanks to the heavy feeding of his cows, were really quite good.

Rex was, in many ways, the outstanding success of his farming generation. A Bluecoat boy determined on farming, he left school at 16, refused to go to college and emigrated to Canada where he farmed, unsuccessfully by his account, for several years in British Columbia. Returning to England, he had a small farm in Sussex and a contractor's business, neither of which in the circumstances of the times flourished. He then came in contact with John Hamilton, a Scottish cattle dealer and speculator in land, who owned a large farm in Chute. It had been purchased some years previously and he was stuck with it. Rex had, he told me, £50 net of debt. Hamilton sold him the farm which was mortgaged and advanced him the balance of the mortgage; he also sold him a herd of cows on hire purchase. Hosier did the same with the milking bail and a local merchant let him have the feed on tick.

This was about 1929. Rex and his wife Muriel then settled into a wooden hut on part of the farm and began milking the cows. They prospered and by 1935 Rex was running two dairies and looking for a chance of expansion, which he found on a big farm near Petersfield. He had sold half the Chute farm and was trying to find a buyer for the remainder when I came along and rented it off him. He had a very unsympathetic bank manager while I was renting the farm. He told me once that he was desperate for £100 and could I pay him some rent in advance. I had to say that it was no good asking me because I simply had not got it either.

Rex went on to emulate Hosier and certainly surpass him in fame. He was a most articulate exponent of his ideas in his prime and he certainly had a greater following among farmers who were outside the areas where bail farming could be managed. His main interest was the production of milk from grass, and he developed a system of dairying which could be adapted almost anywhere.

It was mostly for permanent pasture. In later years he became obsessed with the system and would recognise no other. His great forte was the collection of statistics, of which he had an office full, with which he would prove his points to his satisfaction.

I remember one particular set of figures which demonstrated how different cowmen had different results. Some of them working with similar cows and on the same farm could produce double the amount of milk that their fellows did. The ordinary simple man like me would have sacked the bad cowman and given his herd to the good one. But Rex would have none of these things, and it seemed to me he would use his records as matters for academic interest only.

In later life he subordinated his farming to his obsession for simple grass husbandry. This was all very well in West Wales, a good grass-growing district where he had a considerable acreage run on these lines. But he also had several thousand acres of Hampshire arable land where he could, with much less difficulty, have made a lot of money with simple arable farming once the times suited arable farming, as they have done over the last thirty years. He could also, I have thought, have given credit to the dairy farmers who had been feeding cows heavily with concentrated feeds all their farming lives and made fortunes doing it. There are almost always two sides to every question; and in farming there are usually two answers to the fundamental problem of making a living on a given piece of land.

But once he was established in a reasonable way I always thought Rex was far more concerned with the development of his theories than he was with making money. At one time he had some 11,000 acres – much of it in Hampshire. While other successes of his generation with similar penurious background used their new-found wealth to develop the life style and possessions of the country gentlemen whose estates they had usurped, Rex was always the simple, interested farmer which he had been in the 40 years I knew him. If anyone in farming can be said to have 'walked with kings, yet kept the common touch', it was he.

That being said, I found him impossible as a landlord – as no doubt he did me as a tenant. He had installed a water supply on the farm which was continually going wrong. When I tried to persuade him to do something about it, he refused to do anything though he should have done under the terms of the lease. After a few years he sold the farm, I transferred my anti-landlord activities to his successor, and Rex and I were friends ever after.

By coming to Tangley I also came under the tutelage, although that is hardly the right word, of Roland Dudley, of Linkenholt a few miles away. Dudley, who had been a friend of my father's, was an engineer who bought the Linkenholt Estate in the early 1930s for a song, mainly for the shooting. At first a neighbour farmed it for him in the old-fashioned way and lost a good deal of money doing so. As a rich man, Dudley thought the cost of this was a cheap way of getting his shooting. He would have probably carried on like this, but he met an American engineer on a skiing holiday who told him about the mechanised farming of the Middle West.

Being a man of action Dudley purchased some of this equipment and began to farm Linkenholt on an all-arable rotation using combine harvesters for the first time in Hampshire and all the clutter of mechanisation. He wasn't just an engineer. He was a farmer's son as well, had a feel for the land, and took the best advice available for seeds and fertilisers.

It wasn't long before Linkenholt became the mecca of a number of young men, anxious to have a go at modern farming, who had the chance of taking over quite big stretches of Hampshire and other arable counties which at that time were going begging. Dudley wasn't popular among the Establishment; or among those, even in farming, who thought all grain should come from America. A. G. Street once said he thought the most miserable end for a farmer was to be hung on Coombe Gibbet facing Linkenholt – 'a howling stockless wilderness' he called it.

Dudley himself rather asked for criticism. He was initially as fanatically anti-livestock as the traditionalists were for it. I remember one year I contacted to purchase his barley straw from the combine. I still used a sweep and a horse-driven elevator and asked Dudley if he would let me leave the horse in a paddock while we were harvesting the straw.

'Only if you put it well out of sight of the road,' he said, 'so that no one can possibly say I have a horse here. And make sure that you scrape up all its manure and take it home with you when you have finished.'

I have to say here that Dudley changed his tune in later years and there was a lot of stock on Linkenholt when he died at a great age.

I wasn't very interested in arable farming for the first year or two at Tangley, but when I came to be involved I learnt a tremendous lot from his example. He poured money down the drain when trying out new machines and new methods. Like Hosier and Rex Paterson, he was always ready to give advice and to answer a question. The fact that so many of us managed to become established in the immediate pre-war years was largely due to the help and the example of those three men.

Hampshire in the mid 1930s was also a lot less feudal than Wiltshire. There were big houses and big estates but the old aristocracy had been replaced since the Great War by a new breed of countryman – the successful city slicker and industrialist who valued the shooting and the fact that London was only an hour or so away by train. Few of these people took much interest in their land.

If someone would come along to farm and keep it so that the shooting was improved, so much the better. A lot of Hampshire changed hands in the mid 1930s for between £10 and £15 an acre. The Wiltshire landowners usually had tenants who had cows and had always paid rents, so the estates had kept together. But arable farming landlords suffered with their tenants and if they had no other sources of income, they suffered very badly.

I was never impressed with the benefits of the landlord and tenant system. Particularly not with the notion that the landlord is an essential partner of the farming set-up, providing the capital equipment – the land, houses and buildings – while the tenant provided the farm stock, etc. Either the landlords were incompetent, in which case the tenants exploited them, or the reverse was the case. I never had, as it happened, a traditional landlord. However, when I observed them in action, I always thought that there was too much personal deference on the part of tenants for a healthy relationship. I did have landlords of course, but they were men with whom I could simply treat on a business basis, pay my rent and carry out my part of the lease. I shall have a good deal to say about this later on.

Coming to Hampshire also altered my farming in a big way. Instead of one bail, I now had two, one on each farm. While I often milked as a relief and to see how the cows were doing, I no longer tied myself to doing it regularly. My staff was now five: two cowmen, two boys to help them, and a tractor driver. There was no set time-off or holidays, they kept milking and caring for their cows twice a day 365 days a year. The tractor driver or I fed the young stock, made the hay, and hauled it to the cows when it was too far from the bail for the cowmen to fetch it themselves.

After a while I took on an old age pensioner, one Fred Beaves and born at Tangley who had worked there as a carter all his life. He worked for me until he was well into his eighties; on his seventieth birthday he showed how they used to carry sacks of corn up the granary steps. Although Fred Beaves never took to tractor-driving, his advice was invaluable when it came to working the rather difficult land. Fred told me that he had left school and started work on his eleventh birthday when his father had been killed in a clay pit on the farm. His mother had brought up 11 children and several of them had done well in brickmaking. I knew some of them, and they were all great big six footers and very long-lived in spite of a very poor start, which had included drinking pond or gutter water. That was all there was available in those days at Tangley; at least for the labourers living under thatch.

The treatment of farm workers by their employers, and by the squirarchy, was pretty dreadful in the nineteenth century and some of this bad treatment even lingered on into the 1930s. There were, of course, exceptions where employers or landlords were paternalistic and just, but even these demanded a certain amount of deference and respect in return. I often noticed that few of the men from the villages who went away to both wars ever reached even N.C.O. rank. The only ones who did were the keepers who form a sort of countryside gestapo. All spirit left the country dwellers when the industrial revolution drew away the brightest spirits. I did know of one man who was sacked by a friend of mine after repeatedly failing to get up in the morning to milk the cows.

'The army,' he told him 'is the only place for you. You will never make a cowman if you can't get up in the morning.'

Some time towards the end of the war my friend, who was a sergeant in the Home Guard, saluted a beribboned Major who marched into his yard as he was going on duty. This was the youth he had sacked ten years before, coming to thank him for giving him the chance.

This expansion of my farming relieved me of day-to-day routine and gave me a chance to expand my interests in a variety of ways. Few farmers then or now did all their own buying and selling, leaving it either to merchants, auctioneers or dealers to do it for them. Being naturally mean I begrudged the merchant his profit, the auctioneer his commission and so on. I liked farming, but just growing things is not farming. The be-all and end-all of it is to make a profit. When times are hard, as they were then, anything saved was well worth having.

One of the then current practices of bail milking was that one should stock up with young heifers and, after two or three calves, sell them to dairymen in the vales when they had grown into big and healthy cows. This meant that every year I had to buy about 60 heifers and sell the same number of cows. The theory behind this, much propagated by Hosier, was that this system reduced depreciation and might also raise a profit on the original cost of the animals. The argument about depreciation was a valid one. In those days the average life of a cow in a dairy herd was 2 to 3 years. These Irish heifers might not give a lot of milk in their first and second lactation, but they usually kept alive and improved enormously.

The parts of County Cork where the cattle came from is poor country; many of the heifers were stunted by simple malnutrition. But, because of the policy of the Irish Government, they were very well bred. Ireland, even under British rule, had had bull licensing which restricted bulls used to pedigrees of known performance both for meat and milk. The change from the poor to good land, particularly the chalk hills, did the rest. In those days very large number of Irish cattle used to be sent for sale in England, and some very colourful characters used to trade in them. Buying off them was a matter of *caveat emptor* and bidding was necessarily restrained in consequence. Some of the dealers used to sell through local auctioneers, and this did give the trade a certain respectability. But there were other sales where the Irishman stood with his cattle in the market, and dealt with each customer separately.

I remember one dealer, almost a stage character of an Irishman called Janey Lee, who used to operate in Bristol Market. I asked him once if the bunch of Cork heifers he was offering were in calf or barren.

'I will guarantee them whichever you like, sorr,' he said and asked me £20 a head for cattle which I subsequently bought for £11.

At Bristol Market the Irishmen tried to sell out by about 6 o'clock, so that they could go home on the Fishguard boat. Otherwise the cattle would have to remain in the lairages until the following week or be sent to another British market. The buying technique was to look round in the morning, and then sit on the fence until about 5 o'clock, having made a bid which you thought the vendor would have to take eventually. If he didn't, it was too bad, and there was always another day.

I then decided to go to Ireland to buy them there for myself. The return fare from Paddington to Cork was, I remember £2 10s (£2.50) and this included a berth on the boat. From Cork I went to Youghal by train and there I was picked up by George Tyner a farmer and dealer, who was employed as an agent by a number of farmers in Britain. George was in a difficult situation at that time – an old-time Loyalist and Protestant in a strongly Republican area.

An agent was essential for any stranger buying cattle in Ireland in those days. Every morning we rose at about 3 a.m. to reach the fair in question at about 5.00. The streets were crowded with cattle and were in total darkness except for the occasional lamp; it was almost impossible to see the cattle clearly. Each little bunch was guarded by its owner, sometimes there were only one or two. Nevertheless much of the business was done before daylight.

Dealing was oriental in character, both sides starting poles apart and then gradually getting closer with much shaking of hands and extravagant speech. When a deal was struck, Tyner gave the vendor a sale note and said that he would see his cattle outside the bank at 12 o'clock when he would pay for them. The vendor was expected to load them as well. The reason for delayed payment was simply to check the cattle in daylight, but only an obvious blemish would justify rejection.

The prices of those days were extraordinarily low in any case. In Ireland they were abysmal. This was due to what was called the economic war. When the Irish landlords had been expropriated and the land sold to the tenants, the landlords' compensation, which was known as the Annuities, was placed as a charge on the farms and the Irish Government was supposed to pay them to Britain. During the 1930s, the Irish Government refused to pay these, although it was still collecting the money from the farmers. In retaliation the British Government imposed duties on the cattle exports which together with the freight, meant that a beast worth £18 in England was only worth about £9 in Ireland – a 100 per cent levy for the more mature cattle.

This led to an interesting financial arrangement. The railway gave me a credit account for the freight and the duty which the railway also paid. This could amount to half the value of the cattle I had bought. The account was only rendered in the third week of the month after the cattle were delivered, so some of them would be half financed by the railway for six or seven weeks. This was most useful as I had started to deal in these cattle in a small way, importing them, keeping those I liked and selling on the remainder. During the years up to the war I gradually expanded this side of my business, and was buying and selling cattle in Salisbury and Shaftesbury markets on an increasing scale.

I don't think dealers then were particularly worthwhile members of farming society, but they did help to make a trade and as I ran a fairly large head of stock as my pastures got better, there was always some business to be done. In this I was different from the run of dealers, who generally had little in the way of farms. Many of them did the job for the sake of the gambling element. Few of them have died rich men. There were however some very large-scale dealers, the wholesalers of the industry, and some of them achieved a very powerful position on some markets. This meant that no one bid against them, or else the main dealer or one of his hangers-on would run up the rash bidder, either to stop him or to make him pay too much. Once the newcomer had conformed, he would be made a member of the ring. One man would buy, and then the cattle, sheep, etc. would be knocked out afterwards.

Alternatively the dealers would offer a farmer who had cattle to sell, a deal to stand in. That is they would guarantee the farmer a price for his stock before the sale, and then bid on them as they came into the ring. If they made a profit on the deal, this would be halved with the farmer; but of course if they didn't reach the dealer's price, the farmer would be paid all the same. These practices still exist, but can really only operate in small markets when the trade is bad. Over recent years there has been a concentration of markets into large units, and dealers' rings are very difficult to set up in these conditions.

Although I have indulged in these practices in the past, I found that independence was the best policy. The big dealers would tolerate my minor activities and I was just as keen to buy cheap and sell dear as they were. After the war I started to buy sheep on commission for farmers unable to travel to the sales, but commission-buying is a respectable business. When I got tired of giving my customers credit, I handed this business over to my eldest son.

The benefit to me of my dealing activities was that first of all, it gave me a thorough grounding in judging livestock, their health, and general value at a glance, as well as a fair knowledge of the dirty tricks that almost every one in farming can be capable of, and an appreciation of the use of money. I doubt though if I made as much at it as I would have done if I had stayed at home and cultivated my farm.

This I didn't do. I had two dairies going quite well, I relief milked, but was not tied down, and the rest of the farm ran itself. Although my bank account seldom improved, and was just often around the overdraft limit, I always had a few more cattle about, and it did seem to me that stock meant money. Anyway interest rates were about 4 per cent.

This developed into a life style which I have kept up ever since. I have always employed good men and left them to do the work with as little interference as possible. If I hadn't gone dealing or taken up any of my subsequent activities, I should either have become bored or got into mischief. The real secret of large-scale farm management is to be on call when things get really difficult, to encourage the men, and to see that their requirements are always ready on time.

10. The Docks at Tangley

The next two or three years was a period of consolidation. The pastures gradually improved with heavy stocking and the purchase of hay for winter feeding. The hay was of very poor quality by modern standards; all of it had been made too late and with the seeds already formed. But the seeds shook out as the hay was fed to the cattle – I kept no sheep then – they trod them in and eventually a better sward ensued. That is if the hay had been made of decent grasses. Often it also contained weeds and these added to those already on the farm made some of the fields a pleasant sight for the tourist, but a nightmare for the farmer. Tangley was already famous for its docks. A favourite joke in Andover was to send a newcomer bicycling to Tangley to look at the ships in the docks. The dock, a deep-rooted plant, was encouraged by the deep clay loam which overlay much of the farm.

The docks and other weeds originated from what were called rickyards, where the crops had been stored in stacks until thrashed out in the winter. I had no corn at that time, but the weeds left by my predecessors grew luxuriantly in these places. As there was a rickyard in every field, every field was infected.

From time to time I made a lot of hay, and much of it was swept to the rick by two income tax inspectors from Essex who used to spend their holidays camped on the farm. Sometimes the ricks heated quite badly, and on one occasion we cut a passage right through the rick so that the heat would escape. A bull broke into the rick one day to shelter from the flies, stuck there, and died.

Feeding hay was quite laborious. It had to be cut out of the rick with a hay knife, a two-handed tool rather like a peat spade of giant size, which was very hard work to use. If the hay had to be taken any distance, it would be tied into trusses. I often cut the hay out myself when at home or had it trussed by an elderly specialist in the art for about 7s 6d (37½p) a ton of forty trusses.

Haymaking was always something of a ritual, and began by sending a boy, cost 10s (50p) a week, to a retired shepherd who lived in the village for a weather forecast. This worthy would look at the sun, wind, remember if the moon came in on its back or its side, and would come up with a forecast which would match those of today, at least in the short term. During the war, when weather forecasts were classified, we made the fullest use of George Cook and would not dream of starting any operation dependent on the weather without his blessing.

George had been a drover at one time, and was certain he had met the ghost of Chute Causeway, which lay a mile or two to the north. One evening, he told us, he had been walking home from Hungerford after delivering some cattle. Coming to the foot of Conholt Hill, he had felt almost too exhausted to walk to the top before the last drop down to Tangley. All of a sudden a deep voice said 'Give me your arm'. He felt a strong grip on his left arm and was literally carried at a brisk pace to the top. There he looked around to thank the Good Samaritan, only to find that there was no one there, but his dog was cowering at the road side – a sure sign of the supernatural.

I spent many nights on and around Chute Causeway during the war on Home Guard duty, but never saw anything like a ghost. However, none of the locals would patrol there single-handed. The basis of the story is that, during the Great Plague, many of the villagers of Vernham Dean were infected by refugees from London. No one would care for them, except the then vicar of

Chute. He took to going down to Vernham at night so that his parishioners would not know that he might infect them. The story goes that he himself died of the plague, but has kept to his journey between the two villages ever since.

Certainly the countryside between Andover and the Bath Road was pretty uninhabited in those days; thousands of acres were literally lying derelict and there were many empty houses and farms. The reason for this was quite simply lack of water. There were dew ponds on the tops of the hills which were supposed to be filled by the condensation of fogs and the rain. They were all made to a pattern with steep sides cut in the chalk, lined with clay and straw, and faced again with chalk. Some of them were still holding water when I came to the district but how much came from the dews and how much from the normal rainfall I don't know. I think their great asset was that there was little evaporation on the hill tops, because the steep sides kept the sun off the surface.

The sheep, when they were used to water, drank very little except when on a dry feed like hay. Sheep on grass or roots will go for weeks without drinking at all. Some of the newcomers tried to risk keeping cows on land only watered by these ponds, and found that they hardly lasted any time. On Tangley there were a number of ponds made on the same pattern, but fed by run-off from the roads. These did very well as long as the cattle were not allowed into the water, in which case they used to damage the sides and cause leaks.

For the first two years I rented Tangley under the 11 month lease principle, but in fact never vacated between terms. After a couple of years, Mr Merceron (my landlord) told me that if I wanted to go on farming, I would have to sign a proper annual tenancy and be responsible for the Schedule B income tax, which then amounted to £11 a year. This was a stroke of luck. I had never felt happy with the 11 month tenancy and would not have undertaken it without the other farm on which I had a proper lease. Merceron was in many respects an extraordinary man. He had inherited this lovely estate of about 1,200 acres; but after World War I, he had got rid of all his tenants and pulled down 27 cottages and houses for no very good reason but that he felt he would like to. He had farmed the place himself for some years, but for two years before I arrived he had done nothing at all but make hay. All the gates and fences were well kept up.

He himself lived in a 40-roomed mansion with, at the time of my meeting him, no more than one servant. His four daughters had all been presented at Court and then, without any other training, turned out to make their own living. Much of his income came, he told me, from rack-rented property in London's East End, and when this was blitzed he claimed that all his income had gone. His wife must have had a difficult time with him, as undoubtedly did his daughters, but I always found him perfectly straight and a man of his word. If he said he would do anything he would and if he refused, which was often, he was equally adamant. A friend of mine, who was an estate agent of great experience, told me once that the ownership of English land went to people's heads, and that landowners were never normal. As I now own a fair chunk of it myself, I am sure he was right.

In those inter-war years, tenant farmers were definitely ranked fairly low in the rural social scale, at any rate in Hampshire. Shooting was far more important than farming, and any tenant who failed to grow sufficient roots for the pheasants to hide in, or who cut his hedges too low, would run the risk of losing his farm. For us to shoot a pheasant was, of course, out of the question and the reason why I and so many of my contemporaries never learnt to shoot high-flying birds was simply because any that we shot had to be below hedge level, out of the sight of the keeper. At that time, there were thousands of rabbits and farmers had the right to kill them by any means including shooting. Rabbits were a never-ending menace.

So were cattle diseases. There were two in particular – tuberculosis and contagious abortion. There were always cows going wrong and literally wasting away, particularly as they grew older; this was one reason for keeping the herd young. More insiduous however, was contagious abortion. One year I had 80 per cent of an autumn calving dairy herd abort in July and August; that made a real mess of my budget as they only gave a fraction of milk they should have done. Fortunately I was able to mitigate the worst effects of this attack by buying in more heifers.

At that time there was absolutely nothing one could do about abortion. There was a whole host of quack remedies on the market, but it wasn't until vaccines came along that the disease was kept under control. One supposed cure was to have a goat run with the cows and I provided my herd with a billy goat which I bought, I remember, for 5s (25p). He was a home-loving goat which could eat his way out of any field and was always at the back door trying to get in. I eventually had to give him away after he got into the church at Harvest Festival and ate the offerings and decorations.

Once I had secured a lease, I persuaded my landlord to allow me to plough some bad pastures to reseed. In those days, every lease contained penalties for unauthorised ploughing up; landlords knew that they could never let land which was simply arable. This was a new departure for me as I had done no ploughing for three years, but the Wheat Quota Act provided a fair price for wheat and there was a small guarantee for barley and oats as well.

The system of husbandry was quite elementary. I ploughed the selected field during the late spring and kept the ground worked during the whole of the summer in order to kill the weeds. In September I would broadcast and harrow or plough in wheat and generally found that, after such a summer fallow, I got a very useful crop for those days. Then I would grow a crop of kale for the cows and then spring barley or oats which would be undersown to grass again. Barley was quite a gamble because the varieties were not very good, and maltsters were very particular; but it was generally held that grass needed a cover crop. The real weakness of any system of spring sowing was charlock – a sort of wild turnip. This has an oil-based seed which will last for years in the ground. Most of the fields I ploughed up, while not growing charlock with the winter wheat, became smothered with charlock during the subsequent cropping.

After I had secured my base with this lease I began to expand. I was offered the keep on a 100 acre farm next door which belonged to a retired schoolmaster. Keep was an overrated word, as there was little chance of keeping any cattle on the vestigial pasture, and in any case there was no water on this particular farm. Some of the land was no more than bushes and rabbits, and the pastures themselves were largely couch. I paid the owner, Mr Ponting, £25 a year for the farm on the condition that I could plough and crop it as I wished. The land is not particularly poor, being strong clay and flint, but can be difficult to work in a bad time, particularly in a wet spring.

I fallowed most of the farm, but some of it was either too covered in scrub or too stony to do much with, so I simply left it at the time. However, on the 60 acres ploughed. I grew rather more than 1 ton of wheat per acre which was quite reasonable for those days. Greatly daring, I replanted it again with wheat on the theory that, although it was out of rotation, the average return after two years would be better than growing barley or oats.

This was a mistake. The second crop was very poor and would have been better in an oat crop. I realised too late, that, although the oat price was not very good, oat straw would always sell quite well, or I could feed it to my own cattle. It is a strange thing that, in those days, no one would dream of feeding barley straw to cattle and oat straw would often make almost as much as hay. I believe the reason for this was that, when harvesting with a binder, oats were always cut well before they were ripe, and so made a sort of hay. Today when they are cut ripe for combining, the straw is no better than sticks. Anyway the second wheat crop was damaged by mysterious round patches where the ears were white: 'Take All' of course.

Mr Ponting died early in 1938 and his Executors put the farm on the market. The actual size was 90 acres, plus about 10 rented. There was a thatched house and a range of pretty depressing buildings. I attended the sale and saw it run up by the auctioneer to £950 with no apparent bid. He then asked me to meet him on the farm and tried to persuade me to buy it myself. I bid him £825 and to my surprise and consternation he accepted the offer. I say consternation because no Cherrington had ever owned any land before, and the general attitude among farmers was that only a fool would so risk his capital. In my case this was not even my capital, but Ronie's nestegg plus an overdraft.

However, I was stuck with the house, which was without any conveniences at all, and was thatched into the bargain. Then Hitler came to my aid. During the run-up to the Munich crisis, I was able to sell the house and a paddock for £575. So for £250 we were able to buy some 90 acres of moderate land, or £2.75 an acre. There was one 14-acre field on which I could never grow a crop; in the late 1950s, we sold this for chicken houses for about £3,000. This money, on my father's advice, I invested in shares and it has never really grown in value. If I had still kept the useless field, today it would be worth the best part of £25,000 and I still doubt if it would grow anything.

I didn't have a combine harvester. There were only two or three in Hampshire, and they were not thought very much of by the orthodox. I purchased a second-hand threshing machine for £15 delivered, and ricked the sheaves in the old-fashioned way, threshing them out in the winter. I always liked threshing – particularly cutting the bonds of the sheaves and letting them fan down on to the drum, Threshing was fairly labour-intensive. It required two men to pitch off the corn rick, one feeder, me, two men on the straw rick, and one to clear up the dust and cavings. This was usually a job for a boy or an old man. We also needed a strong man to take off the sacks. I used to pick up an odd selection of villagers, road men and smallholders, who would come and give me a few days work, and my cowmen would help out as well. The output was about 7 to 10 tons of grain in a long day, about an hour's work for a modern combine.

Wages were low. At the outbreak of war, a head cowman would be getting £2 10s (£2.50) a week and his house with a couple of pints of milk a day. For this he would milk the cows and feed them every day of the year except for a week's holiday. A youth on the same bail would get £1 15s (£1.75) for a seven-day week; while a day labourer or tractor driver would receive a basic wage of about the same, but including a house. This would be for a 50 hour week. By this time public holidays were holidays; if they had to work during them, they were paid overtime.

Milk sold used to average about 1s 2d (6p) a gallon and wheat £10 per ton under the subsidy scheme. It is interesting to compare these figures with those for today. A cowman's wage would be at least £80 for the same responsibility, a 35 fold increase over forty years. The price of milk, however, is still only eight times what it was then. Wheat at the time of writing is about £80 per tonne, a rise of similar proportion while the wage of a tractor driver has risen to around £55 for 40 hours, 30 times the 1939 wages. The purchased dairy ration I gave my cows cost, I remember, £6.25 per ton delivered when war broke out. Today the cost of a similar ration is about £110 or 17.5 times dearer.

Farming was profitable and is still profitable today, simply because productivity per man per acre and per livestock unit, has bounded ahead over the years and is still increasing. While increased yields have played their part, they have shown nothing like the dramatic increase in output per man. Average grain yields are about twice the pre-war level, as are those of dairy cows. The productivity of grassland has been greatly increased so that now a cow needs only $1\frac{1}{2}$ acres or less for maintenance instead of the traditional 3 acres.

When I first came to Tangley, the whole farm was infested with docks and ragwort. The docks spread outwards from where the corn used to be ricked, and persisted both in pasture and arable. It was possible to check them temporarily by chopping off the heads and so preventing seeding. But complete destruction was only effected by pulling them out by the roots and by making sure, as does a good dentist, that no matter how deep they had grown, no fragment of root remained.

The persistence of dock roots is legendary. Stan Nokes once told me that his father had taken a dock root home, dried it in the oven for a week, and then carved a pipe out of it, which he smoked for a year or more. Then the pipe had broken, and he threw it into the garden where it immediately took root and grew again. If the men had nothing to do, I occasionally put them on to dock pulling with the help of a special fork with two closely spaced tines. This made little impression and until the right chemicals became available I had to live with them.

They were particularly troublesome in hay. They spoiled the sample and of course reseeded every time the hay was fed out in the fields. This was a nuisance – but at least the docks weren't poisonous. Ragwort is, and hay full of ragwort is a very dangerous feed, especially to hungry cattle. This is an interesting point, because cattle will seldom eat growing ragwort, probably because of its bitter taste. However, once it was dry as in hay, or even sometimes if it had been cut and left lying in the field, they would eat it.

At certain times in the late summer, the ragwort used to be infested with a browny yellow caterpillar, and these used to kill off the plant. I used to leave them uncut so that the caterpillar could do its worse. This used to bring the Weeds Officer, which every county employed, down on my neck with orders to cut the weeds. I don't know if Weed Officers still exist. If they do, they could be well employed dealing with some of the motorway verges.

Ragwort was not much trouble in arable land and could be killed by a summer fallow. But land is so dear now that no one under 50 probably knows what this actually entailed. Basically one had to keep working the field throughout the summer so that nothing grew. There was a theory that the soil should be kept in lumps the size of a horse's head; this meant that the only tool to be used was a plough, and this had to be used every time the field greened up.

This process was quite good at killing couch grass, particularly on clay soil like mine, but was not very effective on light land which crumbled down into a fine tilth through which the couch rhizomes could travel at an alarming rate. On these soils the fallow was assisted by couch burning. The rhizomes were worked to the surface, raked up by hand, allowed to dry and then burnt. In some cases loads of couch were hauled off to rot. All that this treatment used to do in my opinion, was to weed out the weaker couch rhizomes, and leave behind enough of the survivors to reinfest the fields again.

During the Depression, many farmers used to give up their most infested fields, fence them in, and run cattle or sheep. Couch eventually succumbs to constant treading and pasture fertility, and after some years will disappear. However, as few farmers put fertiliser of any sort on their pastures, this process took a long time. But the couchy pastures, without much top growth, gave ideal conditions for the spread of ragwort so the last state was probably worse than the first. Eventually it was found that in fields where sheep were grazed very hard in the spring, they would eat out the ragwort plants and so kill them. I did this and my ragwort problem disappeared.

A summer fallow was also said to be as good as a coat of dung. That meant that you got a better crop of wheat after it than from any other previous preparation. The real reason for the improvement was quite simply that the land had had a rest for a year. Land will always grow something every year; all one was reaping was two years' natural increase instead of one.

Another type of couch was especially troublesome. This was onion couch which grows from clusters of bulbs just below the surface. It does not spread as fast as ordinary couch, but at that stage, appeared impossible to kill. The only effective treatment advised was to fallow the field in mid-April in a dry spring, when the bulbs were vulnerable. As dry springs in southern England are always scarce, it was pretty hopeless advice. Even if there was one, the subsequent crops were generally a failure. So the treatment could only be undertaken as a last resort. Putting the field to pasture did not make much difference to this plague, and it has endured until very recently. Today all strains of couch grass can be controlled chemically.

Soon after I started farming, I visited Jeallots Hill, the ICI Research Centre near Reading. One of the most interesting experiments was concerned with inducing the growth of thistles by intensive sheep grazing. Tight grazing encouraged them; slack grazing, where the grass was allowed to smother them, reduced their number. This was all very interesting, but no real use in a situation where farms were unfenced and few farmers had the capital to keep sheep just to get rid of their thistles.

Most of us used to cut them, and we were guided by the old rhyme which said:

> Cut them in June, they'll come again soon.
> Cut them in July, they'll surely die.

Of course by July they were often ready to die anyway, having seeded, and were unlikely to reseed again or grow from the root that year. They certainly would not throw up a seedhead. Mr Holder used to have a walking stick with a narrow spade at the end – a thistle spud – with which he used to attack the thistles when walking round the cattle. I had one too, but seldom used it as I always used to go round the cattle in a car if at all possible. In any case, my activities would make little difference to the thistles.

In arable crops it was possible to kill them with hoes, or at least to stop them seeding. On grassland we often used old mowers, which usually would rattle to pieces in a short time, mainly because the knives had not enough to cut. The best tool I ever had was an old power-binder, but that had the disadvantage of leaving the thistles in bunches or in a long swath which would kill the grass underneath. Our hay was usually full of them which made it miserable stuff to handle.

There was a theory, supported by masses of hearsay evidence, that thistles never grow from seed. I used this argument whenever angry neighbours or the Weeds Officer used to complain to me about my thistle-down blowing on to their land. I don't know the truth of it, but there is no doubt that the creeping thistle grew from a ryzome and was most persistent. On arable land, we used a tool called a thistle-bar, which was a narrow metal strip fixed to a horse hoe and drawn along some inches below the surface on the theory that, if you severed the roots deeply enough, they did not grow again. I destroyed several of these tools in my clay and flint land, without causing much damage to the thistles.

In 1939 Jack Stratton ploughed a 100 acre down on his farm at Codford which had last been cultivated some hundred years before. The first crop, which was spring barley, came up smothered in yellow charlock. In fact the yellow charlock fields used to be much admired by pre-war country lovers, and their absence is now a cause of some environmentalist complaint. The most interesting thing about this disaster was that even burial for a century would not destroy the viability of these charlock seeds. It was always said, as a sort of old wives' tale, that charlock seed would stay viable for ever, and here was living proof of it. There was a most ineffective treatment by sulphuric acid for it, but otherwise it just had to be lived with.

We used to try and work our spring corn fields early enough so that a proportion of charlock seed would then germinate and could be killed later by cultivations. All this did was to make sure that the grain crop was usually in too late for a satisfactory yield. The only technique that I found to be successful, was to sow the barley with a combine drill with a fairly high nitrogen fertiliser. This made sure that the barley grew fast enough to smother the charlock, instead of vice versa. Even under these conditions the attenuated charlook managed to set seed so the field would be replanted. In the early sown root crops like potatoes, sugar-beet and mangolds, charlock and other weeds could be kept down by hoeing, using horses, the tractor, or by hand.

The later sown root crops – turnips, rape and kale – were a disaster more often than not. If they were not smothered with weeds, they never grew because of the depredations of the flea beetle or turnip fly. Stan told me that once, when working on a big Wiltshire farm where the arable was three or four miles from the homestead, he arrived with the carter's team – of which he was the boy to lead the horses – in a distant field to sow turnips. The whole team was there to roll, sow, harrow and roll again, but they had forgotten to bring the seed. He asked the head carter if he should slip back to the farm to get it, but was told to 'shut up'.

'We'll plant all the same,' said the carter. 'First crop never grows anyway.'

So all day the team solemnly worked up and down the field, going through the motions of planting turnips except for actually letting any seed into the ground. Three weeks afterwards the farmer said to the carter,

'Fly seems to have got those turnips, you had better sow again.'

So they did. It was common to sow these brassica seeds two or three times before getting a stand.

This was not the only hazard. If you didn't get the fields worked down early, so that the weeds could germinate and be killed before sowing, you were asking for the crop to be smothered by them. Charlock and turnips, etc. are very similar in the growth patterns, and so any move to kill the former usually succeeded in killing the whole crop.

Wheat was not much affected by charlock, which has a summer growth pattern and is killed quite easily by frost. But there were other problems. One was advised to harrow and roll winter wheat in the spring. Some people thought that this killed the weeds, others that it induced the wheat to tiller (form side-shoots) all the better and so throw up more heads. I used to do it religiously because I was told it was the right thing to do by almost everyone. That was until I saw that the harrowing pulled out more plants than the rolling ever replanted, while the weeds were unaffected in any way. In some areas, farmers did the only possible thing and set their men to hoe the crop up and down the drills.

There was another old wives' tale about sowing wheat to the effect that it should be sown in mud, while spring barley needed dust. So in a perfect autumn we used to hang on until the ground became really unworkable and then sow the wheat. It was usually broadcast with a horse-drawn broadcast and then either harrowed or ploughed in. On one occasion, in 1939, I actually broadcast some by hand, because I had no horses then, and the land was too wet for a tractor to pull the broadcast.

If by any chance the wheat made some very good growth through the winter, and would hide a hare at the end of February or in March, no one was particularly happy. It was described as 'winter proud', and was a matter for savage attacks with the harrows. Otherwise it was grazed off with sheep. Winter proud wheat used to go off brown; the real cause was that the soil nitrogen had become exhausted. In the end we learnt that one of the best cures, as for most crop problems, was a one or two hundredweight of nitrogen.

Some farmers, and I was among them, reasoned that deep ploughing would get rid of weeds. The theory was that most weed seeds failed to germinate more than six inches deep. So we bought crawler tractors and semi-digger ploughs, and turned the land over as deep as we could. It made no difference to the weeds. Mainly I think because our predecessors had had the same idea, and had buried their problems for us to dig up again. We once put a water pipe in at Tangley and the spoils which were thrown up by the trencher, produced a crop of wild oats, docks, thistles and everything else in the book which someone had thought to put out of sight. I once engaged a machine called a rototiller, like a gigantic egg whisk, which stirred the soil to the depth of two feet. This dug up lots of stones and weeds, and made the land impossible to crop properly for at least three years.

This deep ploughing had another effect. It brought to the surface soil which had no capacity for growing crops until it had been invaded by the soil bacteria and other ingredients which make for a fertile soil. Of course we did not realise this at the time, and most of us went on ploughing and sowing too deep for a long time. Depths of sowing, like the optimum cultivation techniques, are still imperfectly understood even today. But it now appears that, as long as the surface drainage is right, the fewer cultivations which are performed the better. Especially now there are weedkillers and tools to clean the seed bed and plant the seeds.

However, even if you controlled the weeds and got the seeds planted, there were diseases and bugs to worry about. I have spoken about wireworms elsewhere; they used to attack the plant below the surface. Leatherjackets would eat the growing barley and slugs would massacre the wheat. Even if you got the crop through to harvest you were not free of problems.

Wheat used to be subject to smut or bunt. It was not apparent in the growing crop but, when thrashed out, the grain would split or burst and cover everyone in a stinking sort of soot. The cure for this was to treat the seed with a copper sulphate solution, and strictly adhere to the rules of rotation. Wheat was also very subject to bird damage, particularly as it was coming through after planting. The rooks used to delight in pulling it out by the roots, eating the grain and leaving the rest of the plant lying on the surface. The birds used to follow the drills along and pull up plants for hundreds of yards. Many of the older workers had started life as bird scarers at about

2s (10p) a week. We no longer employed boys for this, but used to shoot at them with a .22 rifle. This was fairly effective.

Even if the rooks could be kept at bay, there were the rabbits. After 25 years of myxomatosis, no one really understands the havoc that rabbits used to cause us, especially in the thickly hedged areas where I farm. Rabbits are constant grazers and eat the plant to the roots. Even if the plant survived this treatment through to the spring, the damaged parts of the field would ripen unevenly and this unevenness would persist until harvest, so producing a poor sample.

For many years I employed one man constantly on rabbit catching. He used to have a month off for driving the lorry at harvest. The rabbits were gassed, snared, ferreted, trapped with gin traps and shot at night in the lights of cars. It was essential to keep at them, even in the summer when they were unsaleable. Professional rabbit trappers were little good. They came on to the farm, caught what they found profitable, and always left a stock for the next year. I spent hundreds of pounds on fencing to keep them out of my fields but it was a losing battle. There were too many game-preserving estates around where the keepers, while looking after their pheasants, did little to keep rabbits down. In fact from my knowledge of him the average gamekeeper is probably the most useless individual in the country scene when it comes to killing rabbits or any other vermin which do no harm to his birds.

It was a great blow to our efforts when they made the gin trap illegal. I agree it is a horribly cruel instrument, but no more so than shooting, snaring, ferreting and other horrors. The war against the rabbit was a battle for survival and had we not persisted, the rabbit would have taken over many areas. It is always amazing to me that we tolerate all these cruelties to rats without a word against them. Probably the works of Beatrix Potter have a good deal to do with the sentimental attachment to the bunny.

Just when the rabbits were winning, as they almost did in Australia, salvation was at hand. In the early days of 1953, I believe it was, a French professor inoculated some rabbits in his garden with a strain of myxomatosis, a South American disease of extraordinary virulence. These rabbits escaped and before long the disease had almost wiped out the French rabbit population. Some public benefactor picked up a diseased rabbit in France and let it loose in Kent. The Ministry, after some public outcry, tried to contain it and an order was passed making it illegal to spread the disease. Fortunately this was disregarded and rabbit problems have been almost a thing of the past for a quarter of a century.

Every now and then there is a scare that the rabbits have developed an immunity to the disease and occasionally they become quite thick in some areas, but never on the old scale. Sooner or later, somehow or other, the disease reappears and wipes out the colony; the infestation lasts only a few weeks or months. I believe it to be still largely fatal to rabbits and those which survive, as some do, are probably those receiving less than lethal infection or rabbits which spend most of their time above ground where they do not pick up the flea, the usual vector. The virus is being actively cultivated in Australia. It still works there, and should it ever die out in Britain and the rabbits return in great strength, I can see no alternative to its deliberate use.

I have elaborated on the rabbit because more than any other factor I believe it was responsible for damage to crops and the productivity of grass. After the first myxomatosis attack, it was claimed – by the Ministry of all people – that the benefit to farming of the destruction of the rabbit was to the order of £50,000,000. It must be many times that today.

To get some idea of the changes which weedkillers and other chemicals have brought, it is worth comparing the yields of the late 1930s with those of the late 1970s. The general norm per acre in Wiltshire and Hampshire used to be considered to be 12 cwts of barley, 12 cwts of oats, and about 1 ton of wheat. These would be for crops sown in a rotational sequence so that no grain crops followed another.

Today the national average is 36 cwts of barley, about the same of oats, and more than 2 tonnes of wheat. Many farmers are growing far heavier crops and few of them would be working in any sort of rotational pattern. This is a fantastic improvement when it is considered that the previous yields had been standard for the best part of a century. Almost everyone claims the credit; the seed breeders, the chemists, and the farmers. The answer, of course, is that they all have a share. But unless you have a weed-free crop, such as is possible today, you can expect no yield; nor can you if rabbits eat it off or rooks remove the seed; nor will it grow unless it has the right fertiliser treatment. Even the strains which we used to work with 40 years ago would have improved in yield had they had today's advantages.

11. The Outbreak of War

The war did not of itself mend the fortunes of British farming, but it did restore its confidence. Like merchant shipping and other depressed industries, farming was needed. Farmers and their men found themselves on the schedule of reserved occupations which meant that, unless they wished to because it was not very easy, they were not liable to join the forces. I have always felt a certain sense of shame that I did not rush to join the colours as soon as war broke out. To share, not only the danger, but also the upheaval of leaving wife, family and farm, a disruption which was extremely hard to contemplate with any detachment.

Some farmers did this, either selling up or leaving their farms to others to look after – all honour to them. I rationalised my attitude with the knowledge that I was farming some 1,000 acres and that my finances were extremely strained – my overdraft was never more than a few pounds off its limit. I had no one I knew who could take the farm on in my place, and, being a tenant, I had no house even for my wife and three children. I was also over thirty and had a health problem, a legacy from the Argentine. I had also at that very time just been refused a non-war risk life policy.

Nevertheless I had the urge to do something and in this chapter it's worth recalling that there was a great deal that could be done. Indeed later in the war every civilian had to do at least 16 hours a week of some sort of service: Air-raid Warden, Home Guard, Observer Corp, Fire Service and so on.

During the phoney war period (1939-40), I joined the Fire Service, and spent many evenings learning different methods of fire drill, running out hoses, setting up auxiliary pumps, demolition and rescue from bombed buildings. But in June 1940 there was a call to form the Local Defence Volunteers, later to become the Home Guard. There lay my duty I felt, and as I had had some quite extensive training one way and another, I did have something to contribute.

The Home Guard has been immortalised in the TV series *Dad's Army*. The picture it gives of bumbling incompetence is largely true. Many of the 'Dad's Army' situations occurred in the units I was attached to, and were probably the only spirit with which to meet the well-trained German Parachutists who were going to float down on us from the skies. Any rational assessment of the situation, would be that there would be an absolute massacre, had they opened fire on us.

The first platoon I joined was in the village of Chute. We were organised into a platoon under a World War I veteran, with other veterans as section commanders. These old soldiers may have been very brave men, but few of them had ever seen actual combat under infantry conditions. Most had been drivers of horse transport or other support tasks. The Commander had been a pilot. We were issued with four rifles to the platoon, which went to the old soldiers with twenty rounds of ammunition. The rest of us had to make do with shotguns and old army revolvers – a surprising number of these turned up having been hidden since the Great War. We drilled on Sundays and in the summer evenings; and at nights we organised road blocks to prevent the circulation of spies.

This activity may not have inconvenienced spies very much, but it certainly improved the local morals. No one could take a woman out for the evening in a car, without running the risk of being held up by a grinning couple of

Home Guards and made to show their identity cards, which were being issued by that time. When the bombing started that summer, patrols were out every night; and until the war ended I was out one or two nights a week on behalf of some service or other.

The biggest excitement came during the night of 15 September 1940 when the Platoon Commander called me out with the news that he had heard that the church bells were being rung, and that the Platoon was to mass on Chute Causeway with all the weapons they could muster. This we did, set up road blocks, and sent out patrols. It was a fine clear night. All around the horizon were the flashes of the air-raids on London, Southampton and Bristol and we could hear aircraft overhead. We were stood down at about 9 a.m., when it turned out that it had been a false alarm. That I think was the nearest we ever came to going into action with the enemy.

I left Chute soon after that and came to live near Andover where I joined the local Home Guard Platoon as Commander, a post I held until 1944 just after D-day. As the danger of invasion receded, so we began to get better equipment. Enlistment in the Home Guard became compulsory, and we used to parade as a fairly military force, marching through Andover whenever the Battalion Commander could arrange it.

The senior officers were mainly ex-World War I and there was a constant conflict between them and the younger ones whose ideas of warfare were different, to say the least of it. I was particularly worried that the spread of ability in my Platoon was so wide. Some members were several degrees below par, while others with training, could have been made into very good soldiers indeed. I determined that, if ever the balloon did go up, I would detach a third of my Platoon into an active section, send another third home, and leave the rest to guard the base.

I suppose other commanders felt the same, but I was

always worrying about the difficulties of fighting in the home village with men with whom I lived and worked. Suppose some Germans had gone into my house. Would I be able to order fire to be opened on them? We had by that time got some quite heavy weaponry. Would the men obey my orders to fire on their own houses?

These issues don't affect a commander fighting away from home with soldiers under him who have no local ties. If I lost men in action, I should have had to come home – if I had survived – and tell their wives and families about it. In fact I had a case of this in the spring of 1944. I had to take my platoon on a battle course, which meant advancing down a valley under fire from fixed guns firing over our heads, with explosions in front of us to simulate gunfire.

During the exercise a senior officer flung a bomb in front of my leading section; after it went off, I was horrified to see one of my men rolling on the ground. The bomb was not supposed to be lethal, but it turned out that part of the fuse mechanism had penetrated his face and injured him quite badly. I went up to the officer concerned, pulled out my revolver which was loaded and said that, if he had killed him he deserved to be shot, and I would do it myself. I got the man away to hospital, went home, and wrote a report on the incident. I asked the battalion to acknowledge it, so that compensation or a pension should be paid if it became necessary. My report was not accepted, so I told the adjutant that I would send the report to the local M.P.

While all this was going on, we were stood to for a day or two after D-Day to repel a possible counter-attack. One of my Platoon, a local farmer, refused to turn out and I requested that he should be prosecuted as one or two Home Guard members had been. This too was refused and I was asked for my resignation. In retrospect this seems all a bit priggish at a time when hundreds, even thousands, were being killed in the real war every day. But, while I would have had no hesitation in ordering my men into action if the need arose, I saw no sense in risking their lives in exercises. The Battalion had already had one man killed and several very badly hurt – mainly through failures in discipline when handling weapons and explosives. I thought then, and still do, that these failures were the fault of the officers for not being strict enough and sensible enough.

I had two very narrow escapes in my own Platoon. On one occasion, I was practising firing hand-grenades from a rifle. This involved screwing a disc on to the bottom of the grenade before fitting it in to the discharger cup on the top of the rifle. The disc, called a gas check, ensured that the grenade could travel up to about 100 yards. The men fitted their grenades into the cups and then came up to me for firing. All went well until one bomb only flew 10 yards and exploded at that distance. We were using 7-second fuses and, as I had ordered everyone to lie down, there were no casualties. I had inspected all the bombs to ensure that they had their gas checks screwed on, but I had not supervised the actual fitting of the bombs to the cups. This character had unscrewed and pocketed his gas check for some reason he could not explain.

On another occasion, I was giving hand-grenade instruction with live bombs, and instructing the men in the drill with the usual rigmarole of taking the bomb in the right hand and so on. One man went through the drill, withdrew the pin, waved his arm, and dropped the bomb with the pin out at the feet of the sergeant and myself. We were throwing the bombs into a pit and, as one man, the sergeant and I threw him into the pit and leapt on top of him. His explanation was that he was left-handed, but had not liked to say so.

By the time I left to join the Observer Corps and

all danger had passed, the Home Guard was at least as well-armed, and some as well-trained as pre-war infantry. My Platoon had sufficient rifles, Sten guns and light machine-guns with all necessary ammunition to sustain quite a campaign. I handed over to my successor a dozen cases of Mills bombs, thousands of rounds and various other weapons, all of which had been kept under my staircase. By that time, the Home Guard had become part of the military bureaucracy with regular adjutants, N.C.O. cadres, and a good deal of pompous ceremonial among the officers. Not at all my cup of tea!

Was the formation of the Home Guard worthwhile? Most certainly, even in country districts. It wasn't simply an army of old men: many were young men in reserved occupations who were highly skilled at their jobs and easily taught the rudiments of infantry training. If the unthinkable had happened and there had been an invasion, many of these could have been incorporated in regular formations with a modicum of training. They would, I am sure, have made a worthwhile contribution.

My reason for going into the Observer Corps was simply because I wanted to do some form of national service until the war was over. I also wanted to avoid being reconscripted into the Home Guard, as I understood there was a danger of that. The Observer Corps had been on duty since about a week before the war and carried on until the Armistice was signed. Its task was the visual plotting of aircraft, and the post which was not far from my home was manned continually round the clock. The stint was 16 hours a week: for this we were actually paid, and received a uniform and some ration coupons.

The work was interesting, and I felt that I was doing some good in the war effort. There were few raids by that time and most of the emergency work was plotting the track of outgoing aircraft and those of casualties returning from raids. One Sunday afternoon we plotted a light aircraft, American, until it landed in a hospital's grounds. Soon afterwards we had a signal asking if anyone had seen this aircraft which had been reported missing. Thanks to our efficiency, two unfortunate American officers found their tea-party with the nurses broken up by military police. I am glad they never found out who was responsible, because a Flying Fortress might have unloaded its bombs on all the Observer Posts it could find.

There is no doubt that the war, in fact and in retrospect, was responsible for a spirit of national unity never experienced before. Unlike World War I, the physical dangers were shared between the troops on the various fronts and their countrymen at home. Enemy aircraft could and did fly over every part of Britain so that there was a chance of a bomb even in the remotest glen or dale. Even to hear a distant plane on a dark night or hear the repetitive thumping of a string of bombs could be pretty fearsome.

The odds of being actually hit outside the towns were infinitesimal, but to see the aircraft passing overhead with those black crosses on made me feel that I was the only target. The bravery of those on constant duty or living in a target area is worthy of the highest praise and shall not be diminished by the fact that they hadn't anywhere else to go for the most part. The shared dangers and the shared sacrifices of food rationing, direction of labour, ruthless taxation and loss of individual freedom were the stimuli of vast political change. Everyone was looking for a better world than the one which existed before 1939 with its unemployment and economic depression. These thoughts were encouraged as an aid to the war effort and there is no doubt that, by and large, more people are better off today than they or their fathers ever had been before. They are better fed, housed, hospitalised and educated than any comparable generation.

In this connection, I remember the shock with which the evacuees descended on the countryside at the start of the war. Ours came from London and Southampton and, however poorly educated and badly housed the poor of the parishes were, these people were an invasion from a different world beyond our understanding. It could be, of course, that the ones who stood out were the problem families who could not stay in their billets or were turned from them as being too ignorant and dirty to be borne. Would these things have even been partially cured had it not been for the catalyst of war?

The war also made for an immediate change in my farming system. To begin with the landlord of my farm at Chute, friend Sykes who had purchased it a couple of years before, gave me notice to quit on 29 September 1940. There was nothing legally wrong with his doing this, but it did seem extraordinary that I could be turned out of the job for which I had been reserved from military service so that someone else could take it on. It was also the second time in five years that this had happened to me.

What was worse was that I discovered that Mr Sykes had been made bankrupt for a large sum not long before, and that on the advice of my solicitor I would be very unlikely to get anything towards my valuation and disturbance compensation of two years' rent. The total of this would have amounted to about £700, a very large sum in those days.

A more immediate problem was that the firm from whom I had contracted to purchase about 100 tons of feeding cake for the cows, suddenly pleaded *force majeure* and cancelled this contract after delivering the first 10 tons. This made me decide to sell one of the herds and use the proceeds to buy the implements which would be needed to run my farming on an arable basis. I was sorry to see the cows go, but at that time there seemed to be no way of keeping both herds – a total of nearly 200 cows in milk –

without a sufficiency of high protein dairy cake. Besides, growing grain might be profitable with the guaranteed prices which were brought in for everything we grew.

At this time the War Agricultural Executive Committees, which were set up in every county, were inducing the owners of derelict land to let their farms. I was offered two of these holdings: one was at Chute Lodge close to home, the other was in Hampshire next to Tangley.

The 'War Ags' were given very formidable powers to direct the occupation and actual farming of the land in the sphere of their responsibilities. Their membership was entirely of farmers and landowners with a core of the old advisory services as bureaucracy. The temptations that were manifest in this situation for various forms of corruption were obvious, but on the whole the Committees dealt fairly with their responsibilities.

The farm at Chute belonged to a Mr Addinsell, the father of Richard Addinsell the composer. He was a wealthy man who had done literally nothing to farm his estate over the years. I had fallen foul of him by appearing, on behalf of Chute village, at a public enquiry to try and get electric power laid on. He objected to this as some of the lines would be going over his land. I was asked to give evidence as to the production of the land in question at that time, which was nil and said as much. He was very angry and wrote me a very stiff letter indeed, underlining the rights of a citizen in a free society.

Anyway, the Wiltshire Committee got in touch with me and asked if I would take on the land; either from them, if they had to take it over, or from him directly. Mr Addinsell was still more aggrieved at this and, saying that he thought he would have more control of me as tenant, agreed to let me the land himself. But, he added, he was not going to trust me to set a fair rent, and left it to the Committee to decide, a course with which I agreed. The rent in the end was fixed at 5s (25p) an acre for 2 years and

10s (50p) thereafter. This looks a low rent, but the Depression was only just beginning to lift from arable farming and rents of this sort were common, even for farms with houses, buildings, etc. On Chute Lodge there was nothing but one small set of buildings.

All the land I took over in 1939–40 had been lying absolutely derelict since the previous war or a few years after. It was growing nothing and there was no water, so there was no chance of putting stock on it. Much of it was very flinty, and so had been the first in the district to go out of cultivation. There were thistles and docks, and there were various strains of the grass weeds, couch grass, and a particularly nasty specimen called onion couch which grew from little bulbs. Then, when we came to plough and plant the fields, charlock appeared. This had been lying in cold storage since the last time it seeded. And, if that was not bad enough, there were wireworms. Presumably they too had been waiting there since the farm had last been farmed and were ready to come to life at the first opportunity.

I found out the hard way. I had planted a 50-acre field with barley in the spring of 1940, and the plants had come up beautifully. Then they began to disappear. In other words the new leaves went brown and died; if I pulled up a plant. I found that it had been severed just below ground-level. There was a theory that the way to counter wireworm was to roll heavily. We rolled, and we rolled, and we rolled, but to no avail; great patches of the field remained obstinately bare.

The theory behind rolling was that wireworms could not move in tightlyset soil, so the aim was to compact the land like concrete. But we soon learned that no roller could really set land as hard as that, and could in fact do harm. On one occasion I rolled just after the barley had been planted. The soil was slightly damp and it dried, and as it dried it set. The barley germinated all right, but a large proportion never managed to penetrate the surface and the field had to be resown. It was known that wireworms would follow the ploughing of grass, but in the case of the worst attacks they seemed to have thrived on land which had had very little growth on it for years.

Eventually we discovered that the best course was to work the land so that it was naturally very firm, not above but below the seed. This entailed ploughing in the autumn, and then doing no more than work down the frost mould on the surface in the spring, so that just enough tilth was made to cover the seed. The modern attempts at direct drilling are no more than a logical development of this husbandry practice. But in those days we had no idea of this, or the machinery with which to do it.

The wireworms were eventually defeated by a seed dressing which is fairly effective; but if the seed bed is ever hollow, plant is still lost in spring barley even if the seed has been dressed. Hollowness is difficult to define but can be recognised by its feel underfoot. If you get a slight sinking feeling, it is hollow – and there is not much that can be done about it.

Apart from the weeds and the wireworms the crops did not look too bad in the summer of 1940; for this the combine drill was largely responsible. This, for the benefit of non-farmers, is a drill which plants the fertiliser alongside the seed. Its use was essential because the newly ploughed land was at the time basically infertile: some of it had probably never been properly fertilised. It would have taken many years rotational farming with fertilisers and stock to have achieved any degree of fertility. So, by placing the fertiliser sufficient for a crop right beside the seed, there was a chance that it would last out through the growing period instead of being diffused through the soil as happened when it was broadcast in the traditional way. I was persuaded to put in a trial of the two methods, and the results were spectacular right until harvest. The combine drilled portions won every time. Today combine drilling is losing favour because it is believed that soil fertility as a whole has increased to the point that it does not show any marked improvement over broadcast fertiliser. My land, however, has not yet reached that happy state.

I already had a threshing machine and binders, but decided that the future lay in combine harvesting. There were few available in 1940. What was more, I had very little money to buy one with. There were a few secondhand ones in the country and I bought one for £200 off a farmer in Oxfordshire. It was an enormous tractor-drawn Massey Harris No. 9, which was very similar to those I had worked with in Argentina less than 10 years before.

I went to tow it home one July evening, taking the delivery tractor belonging to the local implement merchant and its driver. The blackout was on at the time, and as we could not drive home at night, we went to the pub for the evening and then camped in a barn until dawn. During the night the air-raid sirens went off and sometime afterwards bombs fell some way away. I don't know how far off they were but they shook the barn in a most alarming fashion and we did not get much more sleep that night.

We got the machine assembled at the farm, and harvested quite a few acres with it. It had a long platform, which was a nuisance as my land was rather uneven, especially the newly broken fields. Every now and then, the far end of the platform would dip into the soil and gather a good shovelful of flints and earth which travelled along the canvas until it plunged into the threshing drum, unless the platform operator could stop it by putting the whole mechanism out of gear. Often he did not, and we had to stop to clear it up. Using a combine harvester was a three-man job: one to drive the caterpillar tractor, one to look after the controls, and one to tie the bags and let them fall in the fields. Afterwards the bags had to be picked up by hand and taken to the barn. Because of our lack of skill, the state of the land, and the contrariness of the machine which was not really built for our conditions, harvesting with it was particularly frustrating. But fortunately the weather was marvellous.

While we were harvesting one afternoon, the machine had to stop and we heard a rattle overhead. This was one of the daylight raids on local airfields and for a week or two air battles and low-flying aircraft of both sides were overhead. The great thing about the combine was that it made so much noise that we could not hear what was going on and so weren't frightened. If we were frightened, we did not pay attention to what was happening to the harvester; so we had to stop the machine and were still more frightened.

The following year I sold this particular monster, and bought two new ones: a Massey 21 self-propelled and a Minneapolis Moline. The Massey was designed for bulk harvesting and I gradually moved away from sacks to the present system.

The 1940 harvest was fortunately fine and I managed to save most of the grain without the use of a dryer. If it was damp, the advice given was to store it in small sacks and eventually these would dry out without going mouldy. These sacks had to be stood separately so that the air could pass between them; they also had to be turned over every now and then to give the bottoms a chance of drying out. The process certainly worked, and I managed to produce barley acceptable to the maltsters in this way.

However, as my arable acreage expanded, this hit-or-miss arrangement, which was very labour-intensive, had to be modernised. So I purchased a dryer from a farmer near Hertford for £50. It was built entirely of timber, and heated by coke; it was very Heath Robinson in its general concept. I went up with one of my men and took it down. Then we hauled it home and re-erected it as it had been before. The only thing lacking was a thermometer, but I was given one from a physics lab.

To use it, you had first to light the fire – not too difficult a job as there was a good draught from the fan. Then when the thermometer showed about 360°F in the hot air chamber, we started the corn running and it certainly dried. In fact I can say that it never failed to dry whatever was put through it but some of the grains were blackened and charred. I used this for a couple of years and never had any trouble with the germination of the sample however odd it looked. Since then I have used modern sophisticated driers and they have produced endless problems in spite of the temperature being thermostatically controlled. I remember reading once that the Romans – or perhaps it was the Celts – used to store their seed corn by charring it. I can only conclude the charring was the secret.

I would never claim that I was one of the pioneers of modern farming. I was no inventor, but I tried to keep up with what might be called the van of progress, once I was sure it really meant progress. In this it is very easy to be wrong. Fashions and techniques in farming alter cyclically, and the progressive farmer who chases every new idea – especially if it costs money – will most probably come unstuck.

When I started farming 50 years ago cows spent the winter tied up by the neck in cowstalls, Then, during the 1930s, it became the fashion to let them lie out in all weathers. Instead of being fatal to cows, as was thought in my childhood, the winter winds and rain did them nothing but good. Then it was felt that they trod the land too much and so they were brought indoors to lie in covered yards on straw. This was grand for the arable farmer because there was a tremendous demand for the straw for them to lie on from dairy farmers who grew none.

About 10 years ago some one invented what are called cow cubicles, in reality a small stall in which one single cow can lie. Very little different in concept from the old stall where every cow was chained by the neck, except that in a cubicle there are no chains. So all a farmer has to do to keep ahead is nothing until the fashion comes round once more. I am hoping loose housing comes back, and then I will be able to sell some more straw!

It was during the summer of 1940 that I came in contact with the Hampshire War Agricultural Committee. Two farmer members, unknown to me, called one afternoon when I had just started the combine. Neither of them had ever seen one at work before and they were not impressed with what they saw. They were even less impressed when they noticed I was burning the straw for which there was no sale. This was not farming according to them. 'What,' I asked, 'should I do with it? Collect it up and put it in a rick and use it for treading into dung?' I already had enough straw for the few cattle I had inside; the rest were all out and in those days feeding barley straw to cattle was considered useless. They went away grumbling and must have put in a bad report because I had a subsequent visit from a senior official and a more modern-minded farmer member. He agreed that strawburning made sense in the circumstances, and my general running of the farm was approved. This was important because the supply of fertilisers and machinery was limited and without the Committee's approval it was often impossible to get supplies.

The farm was completely mapped, the first time since the Domesday book that this was done in England, and I was directed to grow certain crops for which there was a need. These were mainly potatoes, sugar-beet and later flax. A proportion had to be in grain and the grassland had to carry a certain head of stock.

The Committees could overrule tenancy agreements, particularly those parts which insisted that grassland should not be ploughed. Landlords very often resisted this, claiming that their properties were being depreciated, and there was a famous case in Hampshire on this very point. Rex Paterson, of whom I have written earlier, was for a while the Hampshire AEC's blue-eyed boy. Any spare land available was offered to him and he always took it on until, at one time, he farmed around 11,000 acres. Much of the land he took on was derelict or in poor grass and he naturally ploughed it up. Some influential landlords of his objected to this wholesale ploughing and tried to bring pressure on the AEC to restrict his activities in this field, and it appeared to many that they succeeded.

This affair, which became known as the 'Paterson Case', poisoned the atmosphere of Hampshire farming politics for many years and is not quite forgotten or forgiven yet. I was involved as a Farmers' Union member and it is to the credit of the N.F.U. that eventually a stand was taken – an unpopular one at the time – and the whole thing was exposed to the media, so that the principle of disinterested conduct in public life was sustained.

Many farmers, and landowners too, objected to the powers given to these Committees; but I always felt that at time of war, when lives were being conscripted, property had to be made to work for the cause in the same way and without exception. I supported and even served on one of these bodies; but I resigned as soon as the war was over, although they staggered on for some years afterwards.

It was a fortunate time in which to be farming, although there was not much money to be made, or even kept; Excess Profits Tax saw to that. Although there were some elements of black marketing, particularly in livestock areas, there was very little in Hampshire. If your whole farm is mapped, your cropping agreed and the farm inspected, the opportunities for large-scale fraud are limited.

For the first time since I had been farming, the industry was important. There was the encouragement of available supplies and direction of labour so that crops could be grown. We were encouraged to form clubs and groups in order to improve our systems by discussion and comparison. Petrol was never so scarce but that we could find some for these purposes. Thus, I came to know much of the better farming of the south of England.

Dr Blackman of Oxford University ran some of the first weedkilling experiments on my farm. I forget the material used, but it was yellow, and coloured the crops, the rabbits, the partridges, and everyone connected with it. On one occasion I was talking to the Doctor when he suddenly leapt into his car and drove like a maniac across the field, sounding his horn all the while. It turned out that he had seen one of his students pounding up this powder in a barrel – under this treatment it was likely to explode.

Weeds were the restricting factor in all our farming, particularly with spring-sown crops. Much of the land I had taken over was growing nothing at all and it seemed to me that the last occupier had simply ploughed in the weed-seeds and then gone away.

Potatoes and root crops were much more easily cleaned up with horse or tractor hoes. I used to grow about 50 acres of potatoes very unwillingly. My land is not good for the crop because it is far too stony. Even worse was the fact that the Ministry of Food, which in the end controlled the crop, used supplies like mine, grown in non-traditional areas, as a reserve. This meant that I had to keep them stored until the spring before I was directed to sell them, usually to pig-keepers. I did not lose by this, but I used to grow some very good potatoes and it seemed a shame to cast them before swine.

Potato-growing brought me in contact with Scottish seed potato merchants. These gentlemen had a field day because at that time it was impossible to guarantee a disease-free crop unless the seed had been grown under virus-free conditions in the Scottish climate. Today, of course, there are systemic dressings which enable a proportion of the seed to be grown on low ground. So we were at the mercy of the rapacious Scots and well did they plunder the English. The main concern was size of seed. It was supposed to be small, but the only seed I could buy on one occasion although it was said to be able to pass through a 2 inch riddle, had very obviously never seen one. This consignment weighed three to the pound and would have required 5 or 6 tons to plant an acre, instead of $1\frac{1}{2}$. So I set my men to work slicing the tubers and deducted £1 per ton from the bill.

The reaction was tremendous. Mr Peter Fordyce of Auchterarder, with whom I used to communicate in longhand, expressed himself shocked by my behaviour. In all his long years as a seed merchant, no one had ever treated him in this manner before. Such conduct was unbecoming of one he had believed to be a gentleman. I simply returned his letter scrawling at the bottom I would welcome a summons to the county court and would abide by the judge's decision. I had kept a sample of the seed, and I had a copy of the regulations too. This correspondence went on for some years until, in a final flourish of indignation, he damned me and all my works as an 'ignorant oaf'.

I must say that, except on the matter of seed potatoes for which the Scots were and still are notorious, I have always found that farmers and those who deal with them are remarkably honest. There are few written contracts in the trade – although perhaps there should be – and the only way to prepare yourself for an argument, is to hold both the cash and the goods.

12. Stock Development

Although my basic farm training had been with grazing cattle and sheep, the original foundation of my farm was a herd of cows, milked twice daily and the milk sold through the Milk Marketing Board. This was simply because the dairy cow is the most dependable of all the domestic animals for converting grass into cash. I have never known even a moderate dairy farmer to go bust, and dairy cows brought many farmers through depressions.

In the first place, the cow is a good converter. At the time of writing, a good cow will produce 1,000 gallons of milk worth £500 during the course of the year. If you have a beef cow suckling one calf, she will produce about £250 with difficulty. Grazing and fattening cattle will take more than two years to produce a carcase worth perhaps £300. The only trouble with cows is that they have to be attended to without fail fourteen times a week. I used to find milking an interesting job when I was doing it regularly for many years, and was always ready to milk when someone was ill or on holiday because it gave me an idea how things were going. During the last fifteen years that I kept cows, I had them milked on contract. The cowman got so much per gallon and had to find his own labour, while I held the key of the feed shed. With a good man, it was a first-class system and after one bad start. I had a very good man indeed. George King worked for me for 10 or 12 years and eventually he bought the bungalow I had built for him out of his earnings over three years.

I started dairying with Irish Shorthorn heifers, but

when the war came along and imported cattle were difficult to find, I changed over to breeding my own. We were not very good at cattle breeding in those days. Bulls were chosen by eye, and farmers' eyes have always been taken by beefy types of cattle. There was a strain of Shorthorn in Cumberland which was much in vogue; there were regular sales of these in Salisbury. Needless to say they did not milk very well, but they sold all right as barreners.

Until the war dried up supplies, many of us relied almost entirely on imported feeding stuffs supplemented by hay and grazing. The grazing season was very short in Hampshire and nitrogen was seldom used. Grass appeared on average at the end of April, or even later, and dried up towards mid-summer. Then there might be a brief regrowth in the autumn. Dairying was much under the influence of Professor Boutflour, eventually to become the head of the Royal Agricultural College. Boutflour was entirely dedicated to squeezing more and more milk out of cows by feeding them massive quantities of cake. He proved conclusively that mangolds and other roots were an expensive way of giving water to cows; it came much more easily and cheaply from a tap. He was a first-class lecturer, and farmers and his students believed him. His system suited the times and many prospered by it.

In the 1930s, the biggest hazards were contagious abortion, tuberculosis and mastitis. Contagious abortion could almost ruin one. In one year I lost three parts of one autumn calving dairy with a sudden outbreak in August. The cows did not die, of course, but they milked very badly. There were all sorts of quack medicines about, but it wasn't until the vaccines arrived during the war that it was possible to control it. Every year a small number of cows simply wasted and had to be slaughtered. This was because of tuberculosis and there seemed to be no way of stopping its onslaught. Although notifiable, this did not stop farmers from sending contacts into the markets.

Mastitis was also very prevalent and it was generally thought that milking machines caused it. I doubt if they did. The fault was in putting the cups directly from affected cows on to clean ones, and from not taking enough care when washing the udders. On this latter point I used to argue with the advisers. Most mastitis started with a chill in the udder. I thought that this chill was a direct consequence of cold air blowing on to a recently washed udder. Instead of washing udders, I took to rubbing them clean with a damp cloth. The incidence of mastitis in my herd dropped immediately.

The Ministry had a scheme for Tuberculin Tested Milk and soon after the war started I bought a bunch of empty TT heifers – at £14 a piece I remember – and had my own young heifers tested so that when they came to calve I would have the nucleus of a tested herd. I also had my cows tested and found that I had enough approved cows with the new heifers to keep the herd up to strength. There was a ready sale for the reactors, so I bought a few Ayrshires with the money. Within a few weeks, I got my licence and was receiving a premium of $4d$ ($1\frac{1}{2}$p) a gallon. The herd was tested every six months at that time and, if there were any reactors or doubtfuls, had to be retested at much shorter intervals. It took about three years for the herd to become safely clear, but during that time I never failed to get the premium. During the process, I lost quite a lot of milk because I could not keep the numbers up and had it not been for the better returns from grain growing under wartime prices, I would have had a hard job to survive on milk alone.

Being 'accredited' – that was how TT herds were described – made it essential either to buy from TT sources for replacements, or breed my own. The only safe source of TT stock then was the south of Scotland, and the only breed available in any number was the Ayrshire. There was also a Scottish colony in the south. They supported sales at Reading market and I bought a few cows there. But somehow bought cattle, whether from Reading or later from Castle Douglas or Lanark, never seemed to do as well as those I bred myself. I also bought two or three heifers which were a cross between Friesian and Ayrshire. These did extremely well, and I determined to use a Friesian bull and breed my own.

I bought a goodish young bull at Reading – I forget the breeding but I was for a short time a member of the Friesian society – and turned him out in some parkland I rented with a bunch of dry cows. One summer evening I was looking round them, when the bull came up and started to lean against me. He was a well-grown young bull, so I moved over to a tree and tried to put it between him and me. Foolishly I had no stick with me so, as he refused to stop nuzzling me, I caught hold of his ring and tried to hold him at a distance. This seemed to encourage him further, and he kept pushing me round the tree.

There was no help at all in sight, so keeping a firm hold of the ring I walked backwards to the fence and climbed through it. I don't believe the bull had any spite in him, but if I had stumbled or shown any sign of weakness, he would have probably got me down and could have kneeled on me or gored me. This bull had to go after this. On the bail system, there was no provision for keeping bulls under control. This meant that my existing and, so far, quiet Ayrshire bulls had to be used. My crossbreeding did not start until Artificial Insemination began in 1946.

I have always been extremely fearful of bulls, and in the course of my farming have shot several which got out of hand, including one with a Home Guard rifle through the floor of a trailer. The bull, a very good Ayrshire, was trying to tip it over. A.I. has brought not only well-bred strains of cattle to every farm, but also has saved a great number of lives as well. I am glad to say I was partly the cause of its coming to Hampshire.

In 1945 George Gould, a Southampton veterinary surgeon, called a small committee together to start an Insemination Station at Lyndhurst in the New Forest. I was put on the Committee by the local N.F.U., and instructed to press for the service to be extended to Andover. Hampshire and the Isle of Wight were requested to raise £12,000, of which the Andover share was to be £3,000. The money was raised in a matter of weeks in small amounts and the station has never looked back. The original capital was paid back in a few years, and it has been a most profitable undertaking. As their reward, the committee have the satisfaction of having made a most valuable contribution to Hampshire farming. The success is mainly due to the very high standards of operation insisted on by Gould and the Committee in every department of the enterprise. The secret is that the Committee consisted of selected very able men who were determined to make it succeed. They were not, as in so many co-operatives, the results of democratic selection.

My herd eventually became 100 per cent Friesian and, although still on an outdoor bail after a period being milked in fixed machines, achieved some good results. However, some years ago, my eldest son Rowan took on Chute Lodge where the bail was located, because the land was better suited to wintering cattle. George King retired from milking and, after two or three years, Rowan sold the cows and invested in sheep. I always thought he made a mistake, but you have to farm according to your temperament. I made a great mistake in not insisting that the boys should milk from their earliest years. On the other hand had they been so treated, they might have fled farming in disgust and I should have been left with all the land I had amassed.

l must say I missed the cows. Because the milk cheque

no longer came, I was dependent on a much bigger overdraft to carry me through until the corn or the lambs or the cattle were sold. I like a business when money – real money – is trickling in every week, and there is always something to sell every day of the year. This is rather irrational as there is nothing wrong with an overdraft; but the rates of interest have gone mad these days and they are a very costly luxury, to be done without if possible. This attitude may simply be a reflection of my age.

Once the bail had gone, my cattle numbers dwindled. For some years I had a suckler herd, the progeny of some maiden heifers which had accidentally got in calf. They looked well, but were as wild as hawks and when they were joined by a bunch of Galloway steers which I had bought in an idle moment when I thought they looked cheap, they became absolutely unmanageable. To master them I built a very strong set of yards. Once they were incarcerated there, I called in the lorries and sent the whole lot to Shaftesbury market where some dealer from the Midlands bought them. Sheep. I thought, would suit me better.

13. Sheep Farming

As a grazing animal, I am certain that sheep were quite profitable for some years. Unlike cattle, they will live for most of the year on very short grass indeed. A sheep has a mouth an inch across with which to maintain an animal weighing perhaps 150 lbs at the most. The mouth of any cattle beast is no more than about two inches wide, and with this it has to feed a carcase weighing up to half a ton or even more. Therefore it always stood to reason, my reason that is, that cattle would starve where sheep would live. Sheep are also more productive. A breeding ewe will produce her own live weight in lamb every year. A breeding cow will have a calf which will weigh perhaps half its mother's weight at the end of the year and will require a certain amount of feed as well. Once this simple fact had sunk in, I kept no more breeding cows, wild or tame. The ewe will also fatten her lambs in about 4 months grazing while a cow would take double that time and would need supplementary feeding.

However, there are problems, with sheep. There is an old saying that a sheep's worst enemy is another sheep and this has a great element of truth in it. Land where sheep are intensively grazed becomes what they call sheep sick. The ewes do all right, but the lambs scour and fail to thrive – or at least they did. Until shortly after the war, it was believed that the reason why the lambs did not thrive was because the ewes ate the better grass. It has now been found to be caused by worms of different kinds and they can be coped with.

This is not to say that lambs on old pasture suitably

medicated will do as well as they would on fresh clean pasture. But the drenches used will provide sufficient protection for them to be profitably grazed. Profitable grazing means intensive grazing, because land costs are so high now that it's not worth keeping them unless a good return per acre can be guaranteed.

A good return would be seven to eight lambs an acre from five ewes from birth to sale, either as stores or fat. Some people claim more, but I think that five ewes an acre is a reasonable stocking rate.

Stocking of this intensity means very tight grazing indeed. I learnt what I know of this by visiting the Romney Marsh where traditional sheep-keeping has lasted for several hundred years on very good land. There the grass is never allowed to grow at all, in fact any surplus used to be trimmed off with gang mowers. The sheep kept were Romneys, not a very productive sheep but a breed able to stand these conditions. Most of the sheep grazed were adult sheep, mainly two year old wethers, and the lambs were generally unthrifty. Nevertheless, the fact that sheep thrived on grass as tightly grazed as a lawn, made it probable that, once the parasitic problems could be solved, lambs could also be raised on these same principles. I haven't got it quite right yet, but some of my permanent pastures do it quite well.

The essential thing seems to be to have the land in good heart. The pastures need as much fertility as would grow a good crop of kale or wheat. The grass must never be allowed to grow to any really visible length, particularly this sort of permanent grass. New leys will not stand the tight grazing quite so well, and here a certain laxity in grazing can be allowed. It is also important to match the grazing demand to the pastures growth pattern. My grass usually starts growing in mid-March and is heavily stocked from then on. The quality remains good until flowering time, that is about mid-June. Of course it is not

A Massey Harris 21 – John Cherrington's first self-propelled combine harvester with bulk tanks. No more sacks were used. (MERL)

The Hosier milking bail at Tangley. Brian Young, the cowman, went on to manage a very large dairy herd in Wiltshire. (MERL)

Filling the drill. Combine drilling – putting the seed and fertiliser down the spout together – revolutionised yields in Hampshire and Wiltshire during the 1940s. (MERL)

Drafting lambs in the sheep yards at Tangley. (MERL)

JC arguing amicably with Sir Henry Plumb. John Kenyon, the BBC's Farming Producer for over 20 years, is in the centre of the film crew. (BBC)

Looking smart for the Oxford Farming Conference. (MERL)

The pen is mightier than the sword – at work in his office at Tangley. (Peter Adams)

Fishing on the Test at Leckford, which he enjoyed for more than 20 years. (*Financial Times*)

allowed to flower but some of the goodness seems to go out of it about that time. So after mid-June some of the grazing can be slackened.

By this time in any case a proportion of the lambs will have been sold, particularly those born in late February or early March. A single lamb should weigh 40 lbs dead weight at 11 weeks; the doubles weigh about 35 lbs at 13 weeks. I usually sell them early as fat or else as stores in July. By that time the grass is getting less, harvest is on the way and the ewes can be rested and got into good order for tupping again. The lambs would come to more money if kept longer, but I don't think I could keep such a heavy stocking if I tried to keep the lambs to greater weights. I am convinced that profits come from maximising the number of ewes and hence the lambs.

The capital cost of the ewe is the determining factor in all sheep farming systems. I have always bought ewes or ewe lambs as flock replacements from the hill area. These crossbred sheep not only have what is called hybrid vigour – an indefinable productivity factor – but they also have the benefit of a change from hard to softer conditions. I don't know if it is the change or the hybridity which is of the greatest benefit. There is also a certain quality in much of my clay and flint land which grows sheep faster and better than does the light chalk, however well farmed or fertilised. I noticed it with cattle even before the war.

I have a simple rule of thumb for a ewe's economics: she should not cost more than one year's production of lambs and wool. At the time of writing, a shearling ewe costs from £50 to £60. She will have a job to produce much more than a lamb and a half and £3 worth of wool, say £40. As a consequence, I doubt if sheep are very profitable in low land surroundings, if the present price imbalance continues, they could well be reduced in numbers. Two things would change this: a higher price for lamb or a reduction in ewe prices. These may not come about but, unless they do, I think that most ploughable land in Britain will be completely free of sheep in favour of corn-growing before long; especially if rents continue to rise.

Ewes have been dear before and I have explored various possibilities through the last 30 odd years. Early in the war I kept a few Scotch half-breds – Border Leicester x Cheviot. These are big fine sheep but eat almost as much as a donkey. In any case the Scots asked far too much for them. They were a most popular sheep and English farmers with high profits bought them rather than pay taxes. The Clun also had a very popular phase and were particularly favoured because they could be bred pure at home in England without having to pay the breeders a very big price.

I never really liked the Clun. It was bred out of the Welsh ewe by a Shropshire and was a local breed in the Craven Arms district in my early days. Pure bred lambs were never the match of the crossbred and I have never thought a woolly-headed sheep was very productive. Popularity has made them very dear indeed and the Welsh Border Breeders were clever enough to form a society and persuade some of the English buyers to join the Committee, a form of flattery which brought big dividends.

Another Welsh Border breed is the Kerry which has several derivatives the nearer they crossed with the Welsh. At the end of 1944, I went to a ewe sale at Kington and bought 400 young Kerry-type ewes for £1,000. They were a hardy and milky sheep with lowish lambing percentages, but having been reared in hard conditions, they throve with me. I kept this cross for several years and would have them still had they not seemed to get too expensive.

The next fashion was for what was called the Welsh Half-bred, which was a Welsh mountain ewe crossed with a Border Leicester. These became a very popular cross and I farmed several hundred for quite a time. My main criticism was that they were soft, did not hold their teeth, and did not last as well as they should. I also bred a few from ewes bought off a great Welsh sheep-farmer, George Bennett Evans.

George was a great character. During the Depression, he had built his farm, which consisted of most of the southern slopes of Plynlimmon, into a model unit of its kind. Until he took it on, it had been completely unfenced like most Welsh hill-farms. Unfenced land meant that there was no way of improving the grazing because that brought in the neighbours' sheep: there was no real way of keeping the sheep at home. Once he had a ring fence, George was able to improve his pastures by subdivision; the farm was, and still is, a model of its kind. In fact most of the hill areas of Britain would become much more productive under the same sort of management.

George's good turn came when I was indulging in a flight of fancy which could have cost me dear. Like many prosperous farmers in the early 1950s, I was seized with a pioneering spirit and looked for a farm in Wales. Why Wales? Well, it was the nearest hill area to Hampshire. It could be reached in a day, and the land appeared to be cheap. On the farms I saw it was also not too bad – a sort of sandstone, bracken-infested rabbit warren when it wasn't bog. With the arrogance of post-war prosperity. I and many other English farmers thought we could teach the Welsh a thing or two.

Several friends of mine bought into farms and, while those who clung on made good capital profits when they eventually sold, they all admitted to me that Wales, or at any rate the Welsh mountains, were for the Welsh. But it was still attractive to think of breeding your own ewe lambs and cattle in the mountains, and bringing them back home. Besides there is an attraction for a mechanised arable farmer in the pastoral scene in which I had been brought up.

After being pipped on the post when trying to buy several farms, I found myself offered a hill farm to rent carrying about 1,200 ewes in the hills not far from Aberystwyth. It was an attractive farm with two or three promising fishing lakes, and in the early summer it looked very well. The old farmer had gone blind, but he told me that it had always been self-supporting, and he had only bought two rams on his honeymoon some 30 years before. The rent was nothing, and the valuation for the ewe flock would have been negligible in terms of my overall capital. However, before buying, I did the usual calculations on the back of an envelope. This showed that, at the prices of the day, there would hardly be enough income to pay a shepherd to live there and the other expenses, small though they might have been. The answer was to fence, lime and otherwise improve the land to double or treble the stock.

So I went to consult George – the oracle of all things Welsh. He knew the farm, said it was a good one for some Welsh peasant but that it was unimprovable simply because it could not be fenced. Without fencing you cannot control grazing, and without this there is no possible way of improving land at all. In addition it was very exposed and suffered in a bad winter. Go and look at it in February he advised. It could not be fenced because much of the land was held in common; in any case the landlord would not have agreed, even in the unlikely event of the farmers agreeing to fencing, because he valued his grouse. It was a near thing; I might have lost a lot of money and not had much fun either.

The Welsh ewe at the right price is a good investment, but the trouble is that they don't live very well under conditions of high fertility. They seem to suffer more than most from the milk fever type of diseases and won't readily eat supplementary feed containing calcined magnesite which effectively prevents this. I tried several ways of making them feed, even shutting them up in yards with hay and grain feed but they took little interest. Even though losses approached 15 or 20 per cent, I found them profitable at those prices. However, getting rid of the corpses was so depressing that I did the easy thing and bought no more after a few years.

The Welsh ewe is a very good mother and will usually fatten her lamb, but when crossed with the Border Leicester does present some lambing problems. Having a big head, the lamb, is apt to get stuck, and I had some quite heavy losses. Other breeds of grass sheep have the same milk fever problems but because they will take to supplementary feeds, the losses are not so severe.

As well as breeding sheep, I used to keep what could be called heavy fattening sheep. These were a North Country cross out of the Swaledale by the Teeswater or Wensleydale and were called Mashams for convenience. I used to buy them as lambs in the autumn, winter them, and then sell them to the Ministry of Food the following summer after they had been shorn. It was a fairly foolproof operation. For several years they cost about £4 a head and I used to get £8 or £9 plus £1 for the wool. In fact, they were more profitable than ewes and lambs. Not only did I have some hundreds myself but I also supplied a number of other farms, and all went well until the Ministry of Food handed the trade back to private hands in 1954–5.

Private trade no longer found this class of sheep acceptable. The price slumped alarmingly, so I sold the wethers and kept the ewes to put to ram. The Masham ewe is very hardy but not, I discovered, very milky and would not fatten her lambs too well. This is probably because the Teeswater has not the milk characteristic of the Border Leicester and its variants. I knew though that further north the main crossing sheep was the Blue Faced Leicester which would produce clean-headed lambs of good milking quality. The centre of this trade was Lazonby and about 25 years ago I went to one of the sales and bought a load or two of lambs to see how they did.

This cross, the Grey Face or Mule is a first-class lowland sheep, hardy and long-lived. Many of mine are up to 10 years old; they breed a lot of lambs and will fatten them as well. I believe I brought the first of the cross to Hampshire, and today they are probably the dearest breed in the country. Such is fashion. However, unless the relationship of price to value of output changes materially, I think my flock will be gradually reduced.

It will be a pity because, in many respects, sheep are a very attractive branch of farming. Once the worms and the other ills are controlled, they don't give much trouble – but they are best on cheap land. It is significant that in France, where very large flocks were once kept in the arable areas to provide fertility, they are no longer seen at all. There is still no finer sight than a field of young lambs all thriving in the spring and I shall always keep a few to gladden my eye, if not necessarily my bank manager.

14. Wartime Arable

During the last year of the war I took over the remainder of the Tangley Estate and was farming about 1,100 acres there with another 300 acres at Chute Lodge not far away in Wiltshire. In addition, I had 150 acres from the War Ag at Upton and 100 acres already bought before the war. Some of the land was not very good, and the farm at Upton was very steep and infested with rabbits. However, there was one superlative 8-acre field of light land on this farm; it faced due north. Its quality was demonstrated by the fact that it would grow a magnificent sample of malting barley year after year. This was before the days of chemical tests for nitrogen, when the buyers used to judge by eye, the cut of the grain and the thickness of its skin. My samples from this field always attracted the highest bid. The trouble was that I could never grow anything else to match them, and as the total yield from the field was never more than about 10 tons altogether it did not make much difference to overall sales.

Nevertheless selling malting barley was always fun. I started doing so early in the war, and one very well-known buyer from Somerset, Major Allen, used to come to Andover market every Friday to see what he could find. I learnt a great deal about the mystery from him, and even went so far as to purchase a barley cutter, so that I could see just how the grain was going to cut. The criterion here was the degree of flouriness one could see. Good malting barley cut white, while inferior barley looked almost transparent and was described as 'steely'. It was in fact immature.

Major Allen always used to say that the best barley was that which was thrashed out of a rick in November after it had sweated a little. He tried to persuade me to rick some for him. However, there was a catch to this. He would never name a price for future delivery, every sample would have to be judged on its merits. This is a well-known characteristic of all the buyers I have ever met, so I have always taken the line that I will grow the best crops and stock I can, then let the buyers compete for them.

This is particularly true of malting barley where there is only a market for about 20 per cent of the British crop. What I have found is that quality is highly subjective, and is very rarely rewarded as it should be. When supplies are short there is little difference between the best and the worst, it all fetches a high price. But when the market is full, the premium is again negligible. I don't think there is anything wrong with this. It is one of the facts of trading in basic commodities.

This essential truth took some time to dawn on me, meanwhile I spent a lot of time and money trying to get the best samples and the best livestock for a particular market. In the case of grain, all my endeavours have now been set at nought by the fact that the criteria for both barley and wheat are decided by laboratory tests. Merchants and other buyers are no more than posting boxes for the technicians, and don't really need to know any of the skills learnt by experience.

Besides growing several hundred acres of barley, all with malting in view, I grew wheat, oats, about 100 acres of flax and potatoes, and a little sugarbeet. These last two crops were grown on order from the War Ag. and their harvesting meant the employment of either land-girls or prisoners of war. The land-girls did me the most good. On one occasion they set fire to the straw clamp surrounding the potatoes; the fire brigade came out, damped the fire, and so polluted the potatoes with smoke. I got the insurance and did not have to suffer the frustration of watching the crop rot in the clamp as it so often did as there was no sale for it.

Wartime farming was not particularly profitable nor was it as productive as it is today. Feeding stuffs for cattle were rationed and the techniques of making good quality hay and silage were still in their most rudimentary stages. There were shortages of fertilisers and none of the sprays which now keep weeds out of most of our crops. A yield of 1 ton an acre of all grain was quite good, particularly on my sort of land, but this was well ahead of some pre-war yields, particularly of barley.

I tried to farm on some sort of a ley principle, but until the last year of the war I only had one dairy of cows and some young cattle, and although stocking rates were more extensive in those days, they still only provided about 100 acres of real wheat entry as did the 100 acres of potatoes and flax combined. I could, I suppose, have grown two crops of wheat after a ley, or even three as I do today. But the reported experiences of monoculture with wheat had so frightened me that I would not attempt it. Even now I have a guilty conscience when I follow wheat with wheat.

With barley it has always been different. During the last year or so of the war and for several years afterwards, the so-called barley barons flourished. Barley prices in 1949 were over £40 per ton, a level not reached again for 20 years. The barons were the farmers who were said to spend a month sowing in the spring, a month harvesting in the summer, and the rest of the year was their own to do what they wanted. They were really the most fortunate of men, often having got into their farms by the accident of the slump and found that they were suitable for mechanised barley-growing. Some became very good at it indeed, but their neighbours, who were not so fortunately placed or perhaps not so able, used to decry them. In reality we are all very jealous of men who can make a good living without overworking.

It is quite true that prosperity went to some of their heads, and was the cause of a fair amount of instability in their marriages. They became leaders of local society and took to country pursuits just as wholeheartedly as their predecessors on the land had always done. Historically this has always been the pattern of the arable cycle. Prosperity goes to their heads and they live beyond their means until the slump comes. Then they or their children have to go out of business because they can't adapt to the times, and their farms are taken by the sons of small farmers from the livestock districts or from outside farming altogether. Slump, not prosperity, is always the best catalyst for farm change. I have now seen nearly 40 years of arable prosperity and I wonder if it is going to be permanent, which is against all the experience of history.

Many farmers thought the end of the war would bring a change of this sort. They had the example of the years following World War I when, by 1921, food supplies were again plentiful and the Government of the day suddenly repealed the Corn Production Act which guaranteed prices. This completely destroyed the confidence of farmers for a generation. In some respects the situation in 1945 was worse than in 1921 because taxation, which had been negligible in World War I, was pretty savage in World War II; few farmers had amassed much of a nest-egg. I never really worried about taxation then. It was always better I thought to pay heavy taxes than to be killed in the Forces or become subject to Hitler. Taxation was never so bad as to stop one from farming and, if it left no nest-eggs such as brought some (but not many) of my older neighbours through the Depression, subsequent events showed my optimism to be well founded.

However, in 1945–6 there were certainly signs that the Government wanted to revert, if it could, to the pre-war system of cheap food and no commitment to farming except in the short term. Five or six years of wartime shortage had done little to make the leopard change its spots. In 1946 I attended a meeting of world farmers in London, which eventually formed the International Federation of Agricultural Producers. The main burden of their complaint was that Britain, the main market, was refusing a commitment to any form of price support on an international scale.

But in 1946, we had other things to worry about than politics or prices. The summer of that year was one of the worst that I can remember, and it was followed by one of the coldest winters of all time. It was particularly frustrating because I had never seen growing crops look so heavy before harvest. My harvest started normally in early August and continued miserably until the week before Christmas when we scraped the last of a field of oats from the ground and called it a day. I have a photograph of one of my tractor-drivers combining in late November in a greatcoat and a pair of gloves. Much of the wheat had sprouted: when it went over the drier it malted, as did a great deal of the barley.

I was still using the binder both to get some thatch and for oats which everyone said were impossible to combine because of their habit of ripening unevenly. The sheaves of wheat grew out extremely fast and by the time they were ricked, they appeared to be quite ruined. It used to be said that wheat could be harvested when the water was running out of the bed of the wagon. Well, it did in this case.

On the advice of my foreman, Stan Nokes, once a carter of the old school, we ricked the wheat and the oats in long narrow stacks. There we left them until the following June when we moved in the threshing machine. It looked a pretty awful job and it must have been for the men involved: the whole operation was covered in blue smoke. But the corn in the sack was passable for feed and the oats – a variety called Black Tartarian – were as smooth and shiny and sweet as the seed that was planted. The ricks had eventually dried out.

After that experience I never used a binder again, and to avoid temptation sold my share of the threshing machine. A few years ago I was offered a good threshing machine and suggested to Stan, who was still with me, that I should buy it to make some thatch for the very fast trade that was going on. In a very few words he reminded me of previous experiences and said that at our ages we should know better.

By this time I had a good drier and it was kept fully occupied night and day drying, not only my own grain, but also that of many neighbours who hadn't made the same investment. In fact my drier was working on contract for several years after this until farmers decided that it was better to have one of their own.

The harvest of 1946 was utter misery. Although my land is high and with a lot of flint, it became so saturated that the combine and trailers sank to the axles. I remember well a Sunday, the 15th of September, when the clergy were due to join in simultaneous prayer for fine weather. At 11 a.m. that morning I was towing one combine with a caterpillar tractor and another was pulling a lorry with only about half a load of grain out of the field. It had been dry overhead when we started, but at 11 a.m. sharp there was an enormous clap of thunder, the heavens opened and rain cascaded down. We gave up the uneven struggle and went to the pub.

This season altered several of our practices. It was accepted that, in continuous barley growing, each crop should be accompanied by a green crop undersown. Most used trefoil or Italian ryegrass. In a dry season, it did not grow at all and in a wet one it grew too much. In 1946 it grew and grew and most of the undersown barley crops looked just like fields of Italian or trefoil going to seed. I was faced with several hundred acres of this, and it says a great deal for the performance of the combines of those days, that they were able eventually to salvage some grain from the green mess. Other farmers used rape or kale for the same purpose, that is to provide some green manure. I went to see a friend of mine in Dorset, Ronald Farquharson, on some matter and his crops looked just like fields of kale and rape. He was still using sacks and while he was driving me round in the rain, we occasionally ran into sacks hidden under foliage. I began to laugh which annoyed him considerably. So I had to explain that I was only laughing because I had found someone in an even bigger mess than I was.

But after harvest even worse was to come. It began to snow towards the end of January and the snow and frost persisted until the middle of April. Many of the hill areas were worse hit than Hampshire and the R.A.F. was called in to drop hay to the stock. Losses in the hills were high but coping with the conditions was almost impossible at home. This was because the cold and sunless summer had affected the stock and they had no reserves with which to face a long winter even when adequately fed. Where feed was short, it was almost impossible to keep them going. I relied on a lot of kale for the cows, and this was killed by the first frost, most unusual in the south. Also much of the hay was of poor quality. I had begun to build up sheep stocks as well and relied on winter pasture growth for a great deal of their feed – again a fair certainty in the south.

The result of all this was a very much lower milk cheque and a bad attack of twin lamb disease. This last affliction, which has the symptom of coma in the ewes, usually those in good condition, was thought to be simply a reaction to snow – indeed it was called snow blindness. It has now been found to be entirely nutritional in character, and is caused by a break in the feed. It can to some extent be overcome by increased feed.

Anyway it all combined to produce a pretty miserable spring, poor lambing with many deaths, and a lot of hungry cows waiting for grass that was a long time in coming. To make matters worse, the frost killed some of the new grass and some wheat had to be resown with barley. We started sowing again in mid-April and by the end of that month had about 800 acres planted to barley. The summer was dry and very warm and harvest was early, yields were light and quality good. By September, all the horrors of the previous 12 months had been forgotten. This experience has reinforced my confidence in the British, or at any rate the southern British climate. It has never let me down yet; although there were times in the 1976 drought when I thought the end of the world had come, and the British Isles would turn into a desert.

There have, of course, been bad years since 1946. 1950 was one when yields were poor, prices had begun to crumble and the weather was awful. It was the only year when for some reason my farming actually lost money. The 1962–3 winter was in many respects worse than 1947 but we were better geared up to meet the problems, with improved machinery and plenty of feed.

On Boxing Day 1962, I was having a shoot. During lunch in the barn, one of my neighbours was complaining about having too much hay. It was a miserable day, but actually thawing after some frost, and it looked as though a mild spell was setting in. The upshot was that I bought 100 tons off him for £14 delivered. By the end of the winter, the price had reached nearly £40. Moving all that hay to me, when he could have sold it later for three times the price at home, must have been most annoying to say the least.

The snow had returned by the end of the month and on 29 December, the day my daughter chose to get married, there were about six inches everywhere. We had a marquee in the garden heated, I am glad to say, by a paraffin blower which drowned all the speeches – it was too cold to switch it off. That night it started to snow in earnest and there was a blizzard such as I have never seen before. Some of the guests stayed in the house for the evening and eventually Ronie and I retired to bed, having offered the late-night stayers a bed on the floor. In the end they cleared off in a Landrover. At 5.30 a.m, the next day the Road Surveyor rang me up to ask if my snowplough was on the road yet. Because the wires were down, mine was the only phone he could contact from the farm where he had taken refuge.

I told him to go to bed and promised to start snowploughing at 9 a.m. – early enough on a Sunday morning! By daylight the roads were full and I had to send the milk to the depot in my lorry drawn by a crawler tractor. The marquee was full of snow and remained in the garden until the middle of March when the thaw finally came. If there is a moral in all this, it is that if you have a daughter who insists on a big wedding, make sure that she has it in the summer when the worst that can happen is being drowned in a cloudburst.

15. I Become a Landowner

By the time the war ended, it was possible to see certain changes coming which were bound to affect farming, quite irrespective of what the Government of the day decided. I had been impressed by the weight of money which was frozen in this country by the forced liquidation of American stocks and other overseas assets. Some of this money, I felt sure, would be directed into investment in farming.

The amounts available were tiny as compared with those available to the institutions today; nevertheless they were a great deal more than the average farming family – particularly tenant farmers – had at their disposal. There had already been some rather distressing cases where outsiders had bought into farms and had displaced the tenants. Although these were largely prevented under wartime regulations, it was probable that, once these restrictions were removed, most tenant farmers would be at risk. Due to my previous experiences I was particularly conscious of this; and I determined to do what I could to alter the situation.

It was, of course, debatable as to whether anyone should remain in possession of a farm forever simply on the basis of having once gained a tenancy. But, on the other hand, it is not easy to farm positively on the basis of an annual tenancy or even of a lease for a term of years. Tenancies include not only the farm but also the farmer's home. I had got over this by buying a house in a village in 1940 and farmed several farms from this base until eventually I came to own a farm with a house on it as well.

There were two ways in which to achieve this security. One was by means of the National Farmers' Union and the other by means of direct propaganda through the Press, where I was already becoming quite well known. Getting anything done via the N.F.U. is rather like trying to shift a mountain with a teaspoon: it takes an awful lot of time. I was chairman of my local branch and then a member of the county executive, and used to bring up the same resolution on security of tenure at Andover, and then move it at county level, only to find that it was pigeon-holed at headquarters.

Anyway I persisted year after year and reinforced the argument by selected letters and articles in the *Farmers' Weekly* and *The Farmer and Stockbreeder*. These did not gain much public support but did get a great deal of private backing. Farmers, particularly tenant farmers, were fearful of upsetting their landlords. I was in a slightly different position. I had my own house and, in addition, I rented land from a number of different owners who were unlikely to combine together to get rid of me, although I believe one or two of them might have liked to. I also owned a little land which always is a source of independence.

Nevertheless my position was vulnerable. By 1947 we had five children, four of them boys. If I had lost much land, I would have had a hard job bringing them up on the small acreage I owned and the overdraft which was never far from the limit. In those days rents were between £1 and £2 an acre and the total disturbance compensation I could have got, was at most two years rent plus my valuation.

I had my first chance to buy at the end of the war when a neighbour, who remembered the 1921 Depression very well, called me in and offered me his very good 1,100 acres just outside Andover for £30 an acre or £33,000. I told him I had no money and little prospect of any. Then he said that he would leave £27,000 on mortgage if I could find the balance. At the time I funked it. The interest would have cost £2 an acre and there was a huge house which horrified us. I knew that I could farm Hampshire at £1 per acre under pre-war conditions, but to double the rent in the post-war uncertainties would have been foolhardy in the extreme.

I realise now of course that I should have pledged everything I had in order to make sure of such a good farm. But I was not as foolish as Mr Lillywhite, a farmer of great wealth and experience with four sons. He sold his farm the next year for £40 an acre, and thought he had made a killing. So he had, but of his own sons' prospects, none of them were able to farm on any scale afterwards.

By this lime the Labour government of 1945 was beginning to develop the Agricultural Act and I used all the pressure I could to get security of tenure on the Statute. The N.F.U. was not much help for a start. A number of the leading personalities took the view that there was nothing which could be done to prevent men with brains and money getting hold of land and turning the tenants out. They also felt that the union should not try to prevent it happening. They were independent farmers of the old school, confident in their own security and with good traditional landlords; they could see nothing to grumble about.

However, helped by some very well-publicised cases of what looked like irrational behaviour by some individual landlords in getting rid of their tenants, the union did put the proposal in the forthcoming Bill to give tenants a further measure of security. The question was how much? I was called to headquarters to see Christopher Neville, an ex-President and in charge of this particular facet of the Bill. He said first of all that I was upsetting the land-owning class and that many N.F.U. members were landlords too. However, as a compromise, he had secured an all-party agreement that the measure of compensation should be one and a half years' rent as a minimum, plus another year's rent if it was proved to be an unreasonable eviction.

I told him this was completely unacceptable in the circumstances of the time. And that, far from shutting up about the matter, I would keep pegging away as stridently as I was able. He then read me a lecture on the impossibility of running an organisation like the N.F.U. when members actively disagreed with decisions agreed by a majority of the Council. After this it seemed sensible to approach the politicians directly: there were one or two Labour members who had a knowledge of the subject, and were on the committee looking after the Bill.

In its first draft it looked as though some of my suggestions were going to be realised. Full security was granted on condition that a landlord could get possession for his son and this included an illegitimate son if need be. This seemed a good point to object to, because it was clear that an unscrupulous young man could, with the connivance of the landlord, claim that he was the long-lost family bastard. I made a speech to this effect at the N.F.U. Annual General Meeting, with the result that both sons were removed from the Bill.

The result in some ways was a surprising one. I never thought that full security for a lifetime would be offered, but I was looking for much better compensation. If the landlords had been better led, they could have offered a better compromise than the half-year's rent. On the other hand, my friends and I thought it best to ask for the utmost, in the expectation that we would have to give way to some extent and it was better to have something to give way on.

I don't think the resulting security has done farming any harm, although it may have been instrumental in keeping a lot of people in farms who might have had to get out in favour of richer or more powerful individuals. My experience, both as a tenant and an owner, has always been that I am happier farming my own land. Certainly I have never missed the advice or the investment of a landlord. A farm should be able to provide from its own resources the wherewithal to effect improvements, and I have always found that my land could.

By the time security of tenure was enacted, the need for it had passed in my case. In early 1947, I was farming Tangley under two agreements. The first was an annual tenancy for land I had taken in 1944 and the balance – about 700 acres – was on a lease expiring in 1949. I had offered the landlord a 21-year lease for the whole estate at the daring rent, for those days, of £1.50 an acre with a chance of a rent increase or decrease every seven years. This he refused, gave me notice to quit 400 acres and indicated that he would do the same for the other 700 acres when the lease was up. I was reconciled to losing the 400 acres, when he suddenly offered me the option of giving up most of the leased land immediately, and buying the balance of about 700 acres at about £20 an acre. I was never sure of his reasons for this. After farming some of the land for a while, he sold it to another neighbour.

Finding the money was a problem I gave no thought to at all. The price in the circumstances of giving up the lease was a fair one, particularly as I had to rebuild a dairy. I had no money in the bank, and was no more than £700 below my overdraft limit. It was just at the start of the 1947–8 winter, which meant the cash flow was at its lowest. I paid the solicitor £1,400 for the deposit and went in to tell the bank manager what I had done. He read me quite a lecture on the foolishness of spending money like this without the benefit of his wise advice. I told him not to worry because the Agricultural Mortgage Corporation was letting money out for 60 years at $3\frac{1}{2}$ per cent at that point and I only required a bridging loan. He said that that was nonsense and that anything the A.M.C. can do the bank could do better. They would match the A.M.C. interest rate and

I would only pay interest on the balance which I was borrowing.

Shortly after buying Tangley, I had the good luck to be able to buy most of Chute Lodge, about 550 acres. I already rented 300 acres and was protected by the Act by this time. The whole estate, including a large house and a dozen cottages, was put up for sale by the Home Office which had been running it as an Approved School. It was first sold to a speculator who lotted it up and offered the whole estate for sale. I was told I could buy it for £40,000 or just over, as the market was pretty dead at the time.

My good friend, Major Dick Woolley, who was one of the best estate agents I have ever known, advised me to buy the lot. I was only interested in land and, in any case, I did not have the money to put into house property. Particularly not a historic mansion once belonging to Judge Jeffrey with forty rooms and a leaking roof! I bought the land I was renting for £20 an acre and then noticed that, when the lots were allocated, they included two woods of about 50 acres with no access to a road or other outlet within my boundaries. As such they were unsaleable, and I had them thrown in.

The mansion's parks, totalling 50 acres, were sold separately, and the asking price was £50 an acre. They contained some quite good trees. While the sale was going on – it dragged out for weeks, having failed at auction – I noticed a couple of timber merchants measuring the trees one weekend. This alerted me to the possibilities; I shot off to the agent's offices in Bournemouth early on the Monday morning and bought the parks for the asking price of £2,500. This shows how unreliable some agents are, because shortly afterwards I sold the timber in the parks for more than £4,000.

Also included in the sale was something over 250 acres, plus a house and buildings, which had been the Approved School's Home Farm. I heard through the grapevine that the vendors were going to run me up as they thought I would be almost certain to buy the land because it intermingled with what I had bought. Instead of going to the sale, I arranged for a little-known member of Major Woolley's firm to go and bid on my behalf. I went off to Cornwall to do a B.B.C. Brains Trust for youngsters; I was Chairman of that at that time.

Mr Brown bought the farm for £8,000 so eventually I had over 500 more acres for about £16,000. Timber prices went up and I was able to sell nearly £9,000 worth, financed meanwhile by the bank. By this time my overdraft limit was up to £30,000 and two things happened. Interest rates began to rise and reached the then-incredible level of 4 per cent, and the bank asked if I would consider repaying about 10 per cent a year.

I told the bank manager not to be bloody stupid, and asked him how he thought I could repay with taxation as it still was in the years after the war? I went to the A.M.C. and borrowed the lot at 4 and $4\frac{1}{4}$ per cent over 60 years. I then concocted a marvellous letter to the bank asking for $\frac{1}{2}$ per cent on £30,000 over 59 years – the cost of accepting a bank manager's advice! Needless to say I got nowhere with this.

I did have one more land deal at Chute Lodge. About 100 acres, mainly woodland, had been sold to a timber merchant and this was for sale, after he had done the felling, subject to a planting order. I bought it in the end for £250 and, like a good little boy, offered it straight away to the Forestry Commission on a long lease. The Commission's Acquisition Officer came to see me and I agreed to let him the wood for 200 years for I think 3s (15p) an acre. I insisted that the commission should erect and maintain a sheep-proof fence on my boundaries. This they refused to agree to. I was so incensed by this parsimony that I refused to carry out the deal at all, and cleared 60 acres with the help of a grant. It was a cheap buy altogether.

I was extremely lucky, but many of my neighbours thought I was out of my senses, especially to have borrowed all that money when I would, in future, be liable every time a roof blew off the barn or the house wanted painting. I did have quite a few qualms myself, especially as farming was not particularly good in those days and any surplus cash was spent in keeping the buildings, etc., in condition and generally doing the landlord's share.

It was suggested that I might like to have a lease-back, that is a capitalist would buy the farm and lease it back to me. This would have not only freed me of any debt, but it would also have given me the capital to develop my farming more intensively as a tenant. I resisted this without much difficulty. I had no intention of going back to being a tenant and was, in fact, actively looking for more land to buy as my sons were growing up and showed signs of wanting to farm.

Even my dim wits understood that, while revenue profits were highly taxed, capital profits – i.e. on land – were not taxed at all then. It became apparent during the 1950s that land values were slowly rising, so it was obviously best to own as much land as possible and just farm it well enough to keep it going. This policy kept me extremely short of ready cash. To some extent, I was only able to go in for land-ownership by denying my children what everyone considered to be essential: a good public school education.

I had hated Berkhamsted so much that I felt no inclination to let my children be subject to such an experience if I could help it. My resolve was strengthened in that there was a grammar school in Andover; my four eldest children all passed the eleven plus and went there. This meant that, during these years when I was building my business, I had no school fees to find. Even in those days they took a pretty high proportion of taxed income. I may have been selfishly denying them opportunities in our society, and I certainly upset my parents by this policy; but I hope they haven't suffered. I don't think schools make much difference when the children are living at home. Although it may be hell for the parents, I think day schools are the best.

There is another point worth considering. Our main competitors in the world, France, West Germany and the U.S.A., have almost universal free and competitive education. I am sure that the divisiveness of much of our society is due to the segregation inherent in the public school system. How can the managers, who often come from such a background, understand the problems of those whom they have never met except in confrontation across the bargaining table.

In 1955 I bought 800 acres at Hippenscombe for £20,000. It was a farm which was largely an enormous rabbit warren. Hosier, who owned it at one time, told me he had wintered about 100 cows on it, dry cows at that. It was so steep that only about 100 acres was easily cultivatable. The house was a ruin and there was a very poor cottage. I bought it, I remember, on 29 May and Stan Nokes and I drove over the crops of the vendor, roughly 200 acres, for which I had given £9,000 together with some machinery. They were so sparse, owing to a spring drought, that the car wheels made no tracks on them.

We decided to put on nitrogen and to spray the weeds. Then the heavens opened and there were ten days of heavy thunderstorms and warm weather. Everything went well and we harvested 34 cwts of barley per acre; the highest yield we had recorded up to that date. If it hadn't rained, things would have been very different. I again borrowed most of the money for this at $4\frac{1}{2}$ per cent from the A.M.C.

But my luck in these low interest rates did not hold, and it was entirely my own fault. In 1962 I was actively looking for a farm for my second son, Dan, who was in

Australia after University and urging me to find him one. It would be a bad thing to have him with me or too near his brothers so I looked all over the south of England and was actually last bidder but one for a number of moderate farms. By this time Dan was at home, engaged to a charming New Zealand girl, and was getting more and more frustrated. At length I heard of a coastal farm near Plymouth, 1,100 acres of what I took to be quite good land.

I bid £100,000 against an asking price of £115,000 and as the agent kept ringing me every day, often twice or three times, I thought it would come my way. Then he said he had another customer who would buy half of it. Eventually he asked me to go down to his office on Christmas Eve, meet the other buyer, and complete the deal. This we did and I was quite surprised to see that the other party actually existed. I bought what I considered to be the best half for £47,000.

At that time, the bank suggested I went to the A.M.C., but I said only a lunatic would expect interest rates to remain at their level of $6\frac{1}{4}$ per cent for much longer and I would go to them at $5\frac{1}{2}$ per cent when it happened. I was so wrong in this. Before interest rates fell, I sold this farm to Dan and Chute Lodge to Rowan, my eldest son. They still think father knew something, especially when interest rates hit 16 per cent.

I was, as I said earlier, immensely lucky to have collected all this land without any capital at all to start with. But I wasn't very clever: I could have bought most of Hampshire for less than £15 an acre in 1935 when I went there first. Again I could have bought for less than £30 an acre just after the war, for £50 in the 1950s, and for £100 in the 1960s. But I bought with no thought of capital gain and against the advice of men usually considered shrewd, because it was the only way I thought I could keep farming in the family.

16. A Question of Finance

At this point I think I must make a confession. Through all the years I have been farming, I have never kept a reasonable set of books. I have, of course, produced accounts for the Inspector of Taxes with the help of my accountant. But these are of historical interest only and are far too late to influence my farming.

My reasons for this are certainly not praiseworthy. I know bookkeeping and financial control are the essence of any business. But in one as small as mine is, the total turnover this year is around £200,000. As I vet all accounts and expenses, I feel I have my finger on the pulse. I know how I am doing and don't wish to have to face my mistakes in black and white.

Of course I have been very lucky. I started farming at the absolute bottom of the inter-war Depression, and progress ever since has been always upwards. In 45 years, I have not known anything of a permanent downturn; unlike those unfortunates who came in, say, 1919 or 1920. Undoubtedly one will come one day, when either prices will fall or remain stagnant, and costs go on rising. But my luck has meant that most years there has always been a slight rise in returns to overcome increasing costs. Sometimes the increase has come from higher prices, sometimes from higher yields, or what is now called productivity. But these increases usually kept ahead of costs.

I always knew from my bank account and annual valuation roughly how much I was worth; if I had found my living expenses and kept the farm going as well, I knew things could not be too bad. I simplified this basic accountancy by never owing any money except to the bank. I only once asked a merchant for credit and that was when I was buying a farm and the bank was slightly restive. I took the credit for about three months, and then was happy to pay it off. Apart from the slightly dearer cost of the credit, because of course the merchant would have to borrow the money first, it worried me to have this outside creditor to consider. For several years afterwards I felt obliged to deal with him out of a sense of gratitude.

This last obviously cost me money from time to time and, as soon as my conscience would let me, I reverted to my old system of paying cash and taking discounts wherever I could. I don't think credit from a bank today is particularly cheap, but it must be cheaper than from a merchant.

I don't much care for the modern banking development where agricultural experts, working for the bank, involve themselves in a farmer's whole enterprise. They insist on cash flow forecasts and a lot more information and appear, from what I am told, to take a major part of the farming decisions, but without accepting any of the risks. My view of banks is that they are just like any other trader with something to sell – in this case money. They want security naturally enough for the loan. Therefore they should take the security as collateral, and leave it to the farmer to look after the money and presumably pay it back one day.

I say one day, because banks are supposed to lend on the short term. The manager tells you when he grants the loan that a bank is not allowed to lend long-term and he would like phased reductions. This was always established banking custom, but circumstances alter cases. Even during the Depression, most farmers were good risks. They kept up the service of their loans and the banks carried them. Today farmers are even better risks.

Their assets far exceed their liabilities, and the farming industry as a whole is probably still a net creditor of the banking system. Therefore, whatever a manager tells you about repayment, it is really the last thing he wants. He might be lending you as a farmer £20,000 or so on a more or less permanent basis. But this is a much better bet for him than eight or ten little loans of £2,000 each. Just think of this next time you are in his office.

The only really accurate system of farm accounting is enterprise costing. This means that every man and every machine has his work recorded, costed and attributed to whatever crop or stock product incurs it. It is time-consuming and expensive to do on a largish scale. I have never attempted it. But I have friends who have done so and they have always told me what their researches have revealed. As their farms are comparable to mine, I have used their results to plan my own farming to an extent. Or perhaps I should say to confirm my own judgement.

Then there is the gross margin system, otherwise known as the poor man's guide. It is not as expensive as enterprise costing, but it does give you a general idea as to what is going on. Neither system gives you a real basis of farm planning, because neither attributes overheads to the specific crops being looked at. By overheads I mean machinery, land and maintenance costs, besides all the time when the labour is doing nothing in particular, while waiting for the rain to stop.

As an example many farms are run on the ley system, using sheep and cattle to rest the land from over cropping. Under either of the above systems, the livestock acreage has to carry the same overheads as does the cropping, although the returns are often much lower. I agree that, in fact, these livestock enterprises are probably less profitable than cereal cropping; but if your farm is such that the only reasonable break crop is grass, it would be sensible to set the lower return from the grass against the returns from the arable as these should really be a cost of arable cropping. You don't for instance need a whole stack of machinery and labour for keeping sheep or breeding cows. Indeed there was a time, a few years ago, when a good case could be made for extensive stock farming as an alternative to cereals.

Of course accountancy is easy if you only have one product to sell like milk. All you have to do is maximise production as cheaply as you can, knowing that all your efforts are only directed to one end. It is when you have two or three lines – a mixed farm in fact – that complications and conflicts of interest can start.

Nowadays I have three lines of output: cereals, pigs and sheep. I have cereals because they are quite easy to grow and handle. Sheep because I like them and I have some 200 acres of either unploughable or difficult grassland, and they are more profitable than cattle. I have pigs, also because I like them, and they also fulfil two very important functions. They consume about 400 tonnes of my own barley every year, and they also provide a substantial monthly turnover. I would be the first to admit that turnover does not mean profit of itself. But they are profitable most of the time, sometimes very profitable, sometimes not. But, like a milk cheque, a pig cheque pays most of the current expenses.

I do cost the pigs to this extent. I cost my barley at what I think the average price will be during the season. This means that it is higher in the autumn and perhaps lower in the early summer. From this I work out the cost per tonne of the rations, multiply this by the number of tonnes produced, and there is the feed cost and other incidentals. Then I deduct this from the total sales of pigs.

If I have a good year, I repair buildings, put up new ones and refurbish the equipment. Then I write them off at once. When my accountant says that some buildings must be carried on the accounts, I obey him as far as the Inland Revenue is concerned, but in my own accounts or such accounts as I keep, they have already been accounted for. They are certainly not an asset that could ever be cashed.

Many years ago I was visiting a rather academic farmer and, while looking round, I noticed that half a field of barley was smothered in charlock while the rest was clear. I asked him what had happened, and he told me that his budget restricted the amount he could spend on sprays and this is where the allowance had run out! This, I agree, is an extreme case, but it does underline one of the weaknesses of governing your farming by budgetry means. It is all very well to cost your inputs with so many cultivations, sprays and so on. But you must be flexible. Sometimes you need twice the work to get a seed bed that you do in other seasons. In farming you have to do the best you can, to get the best result, and you cannot budget your costs with any real accuracy.

Nor can you budget your returns. Yields, prices and costs are infinitely variable. I have seen the most elaborate farm budgets which make some extraordinary assumptions about these and forecast a reasonable but narrow profit margin. I do make out rough budgets, based largely on experience; however, they are on a whole farm basis. I take my known cost elements: wages, rents, machinery and other costs from my accounts. I then estimate very conservatively the yield per acre per sheep or per pig on known or estimated prices, and compare the two.

Unless the margin between them is really substantial say 50 per cent of turnover, I scratch my head and start again; anything else is simply not safe to me. I think this is because I am biased by my own likes and dislikes in making any assessment of a farm project. This does colour my calculations so to compensate I try to be over-cautious.

Of course there is another most important point. A farming business is a long-term affair. It is not one to alter or mess about with for short-term considerations on the basis of one or two years' accounts. There is a danger that, if too much attention is given to fine tuning the accounts, one could lose touch with the farming implications of any decision.

For example, almost all my farming life wheat has been – on paper – the most profitable cereal to grow. It yields better and is worth more than any other. But all my training was based on the fact that wheat had to be grown on a rotational basis once every three or four years. I discovered, by intuition if you like, that in my circumstances the other crops in the cereal rotation did not pay as well as wheat. I also found by hard experience, that trying to overdo the number of successive wheat crops was a road to disaster. I eventually took up continuous barley-growing which in the circumstances of the time did seem to work. Although the annual return was well below that from a crop of wheat, the aggregate over four years worked out at more money. Here was a case where an obvious accountancy advantage, had to be subordinate to farming realities.

There are other considerations as well. During the 1960s, cereal-growing was not very profitable. So, about 10 years ago, soon after I had let Rowan have my dairy, I restarted a pigherd. Not because I thought they were particularly profitable, but because they could turn my grain into cash, perhaps a little more cash than otherwise, and give me a weekly cheque as well.

This calculation has been upset by the high grain prices in the E.E.C. system, and probably it would be more profitable to turn the whole farm into a cheaply run grain ranch. But I cannot believe that the advantage given to cereals at the moment can possibly last. It is against all common sense, and so obstinately I am resisting selling the pigs and the sheep, to become a wheat, barley and seeds baron. If I kept and believed in accurate accounting above all else, I would certainly be tempted to try.

Someone told me the other day that it was all very well for me to boast of farming by the seat of my pants, bolstered as I was with 50 years of experience and a farm at cheap historic costs. I am very conscious of my good fortune and would be the first to advise any young man to look to strict accounting and even budgeting to keep on the right lines. But the whole purpose of self-employment is to be able to do exactly as one pleases, and I enjoy my farming and not my office work.

17. Outside Interests

I discovered soon after I started farming on my own that the simplest of farming systems were the best. The more complicated the enterprise, the more time which has to be spent in sorting out the problems that arise every day. If you just keep cows, you only have the cow problem; but if you have half a dozen other lines as well, you have six different problems to face every day. This means that so much time is spent on management that you simply don't have time to think, or enjoy whatever it is you enjoy in life. Once you have a good system that pays, it is best left alone.

Of course, if you are one of those who enjoy the process of management for its own sake, that is fair enough. But management is a much overrated occupation. I also believe it is something which cannot be taught – one either has the aptitude or one hasn't. However, I haven't been as rigid as I should have in my principles. I have never been content with a simple existence and have always complicated the issue to some extent.

The only time I had a really simple system was when I moved from West Knoyle to Tangley and had two outdoor bails, a few store cattle, and nothing else. I had two good cowmen, two boys to assist them, and a tractor-driver to help make hay and feed the dry cattle in the winter. I myself was relief milker, general dog's-body, and supervised the lot. I never minded milking in the mornings, but rather grudged giving up the rest of the day to it, which was entailed by afternoon milking. So I began to look for a profitable use of my time after the morning milking finished – in those days at 7 a.m.

I had plenty of spare land on the two farms which grew a little inferior grass, and I began to stock it with young cattle. I took to hanging around markets and picking up the odd beasts which no one else would buy. I found that, while good-looking cattle always commanded a premium, there was more money to be made out of ugly ones. They usually doubled in value in a year while the good ones appreciated perhaps by 50 per cent. This essential truth, which still holds good today, was one of the best lessons I ever learnt, and made all the hanging around markets worthwhile.

My pre-war system needed little management expertise. The main product, milk, was sold through the Milk Marketing Board. This organisation while marketing milk, has successfully kept farmers in ignorance of the milk market for the last forty odd years. My other sales of grain and lambs did give me a certain amount of practice which has stood me in good stead ever since.

I had started keeping sheep about two years before the war with a flock of Cheviot ewes bought for £1 per head in the Berkshire downs. I ran them near my home at Greens Farm where the pastures, following Rex Paterson's treatment were very good indeed. My reasons for keeping sheep there after the first two years instead of cows were quite simple. Under the pooling system run by the Milk Board, the Western Regions paid something like 1d or 2d ($\frac{1}{2}$p or 1p) gallon less for milk than the Southern Region. Greens Farm was in Wiltshire, part of the West Region, while Tangley was in the Southern. So, as soon as the Tangley pastures were fit for it, I moved the Wiltshire herd there, so gaining an extra 50,000 old pence or about £350 a year – more than enough to pay the cowman's wages with a bit over.

I tried sending the lambs to market, but I found that the buyers were very capricious. Once I had got them in market, I was at the mercy of the trade of the day. In addition, I had to pay the costs of getting them there and the auctioneer's commission as well. So I approached the biggest butcher of the district and tried to get him to buy mine on the farm. He came down one Saturday, handled the lambs, chose what he wanted, and gave me a fair price for those days, about 30s (£1.50) per head.

I developed this home-selling further by being one of the first to sell deadweight to wholesalers. Many farmers are fearful of doing this, but I always found such selling to be the fairest all round. Responsible and businesslike butchers, and most of them are these days, want good stock and I have found that they appreciate deliveries of suitable animals. I always found these buyers went out of their way to show me the type of animal they required and how to judge them on the hoof. I don't believe any particular system of selling can beat the market price of the day by more than a little; but I reckon that selling all my lambs and pigs deadweight, does save me several thousands of pounds a year in expenses and commission alone.

However, you must know something of the trade and its demands, and it is always worthwhile to go to market and study the scene. I was so sure of the rightness of this, that I used to teach my sons how to draw lambs on the hoof from the age of 15 or so. Of course like all boys they knew better than their father, so I took to sending them with a pen or two of their own drawing to Salisbury market where they were graded, with orders to bring back any that failed. They very soon learnt after they had had to bring a few home.

I completely disagree with those who claim that a farmer's job is to produce, and that all the selling should be left to the professionals. This means, I suppose, merchants suitably tamed by being employed by farmers, co-ops and so on. There must be exceptions to this. Milk is a suitable candidate for organised marketing. I would think that grain is too, and I would vote for a compulsory grain board which would buy and sell all grain on grade through a monopoly channel. Wool and hops are obvious, but livestock are so infinitely variable that a complete board would be difficult to run and administer.

My only real principle of management is to put a man in charge of a department and leave him alone. Good men always respond well to this treatment; bad ones fail at it, and will often refuse to take it on. When I was running two bails, I used to go and see the cowmen when the sun was shining, everything in the garden looked lovely, and they had time for a chat. When conditions were atrocious, I used to time my visits with care and then took pains to sympathise and to encourage or lend a hand. I always tried to make them think that any improvements were on their suggestions – they usually were.

My arable work I handed over to Stan Nokes who, when the war started, said he wished to give up cows and go back to the plough which was his first love. He was a natural arable man with an inherent knowledge as to just how many cultivations and of which sort would be needed to produce a seed bed. He never failed through the worst of seasons; at one time, I used to give him a list of 800 or so acres to plant and it was always done right.

I was never much good at arable farming but I generally supervised the livestock and did all the buying and selling. Stan's son, Desmond, now looks after the arable and although without Stan's early training, does understand the confusion of sprays, which are the substitute for husbandry these days, in a way his father and I simply can't. Once the practical work is in good hands and can be dealt with in about half an hour a week, except in busy times, the rest of the time is my own. What to do with it?

A year or two after the war, I was on the forecourt of an Andover garage. When turning my car, I had the misfortune to touch the plate-glass window of the showroom and it shattered. I went into the office to apologise and found the managing director, Walter Lansley, looking at the brochures of an unfamiliar tractor. It was the Ferguson. Harry Ferguson, its inventor, had quarrelled with Henry Ford in America, and had transferred his operations to Britain using the manufacturing capacity of Standard cars.

Harry Ferguson, whom I was never fortunate enough to meet, was a gifted Ulster engineer who invented the hydraulic power system for tractors. He called it the Ferguson system, and it has quite literally revolutionised tractor design since the war. He was even a better salesman/evangelist. Anyone who wished to manufacture or sell any parts of the system had to have an absolute belief in its merits. Until he arrived on the scene, most farm machinery was built with the tolerances and steels of the horse age. Ferguson used only the best.

Anyway, Lansley showed me the brochures and told me he had been approached by Fergusons to market the tractors. Apparently the main dealers already tied to the more established makes like Ford and International had no wish to change, and one of the conditions of the franchise was that competing lines must not be carried. After some talk he invited me to join his firm and to start a subsidiary, Anna Valley Tractors, of which I became managing director.

This was a new departure for me. Previously my only experience of machinery dealing was on the other side of the counter when I was more interested in obtaining discounts than anything else. Frankly I did not think the tractor offered – a small petrol model in the first place – would have much of a future among the large farmers of North Hampshire. I was wrong on two counts. For the first year of our operations, petrol was still rationed and a petrol tractor was the ideal way to get hold of a bit more. The other point was that it was such an extraordinarily useful workhorse, particularly with the hydraulic lift.

This device enabled Rex Paterson to design a buck rake for making silage which was quite one of the most brilliant creations of his fertile brain. The buckrake permitted masses of heavy wet grass for the first time to be picked up and stacked without any need for hand labour. No other tractor at the time had the hydraulic lift, and we had a clear field for several years, during the course of which we sold some 500 new tractors in North Hampshire. Most farms seemed to find a need for at least one.

They were cheap too. The price for several years was just over £300 and the manufacturers gradually built up an impressive collection of matching implements. However, they failed to capture the market completely because they weren't big enough. Ferguson claimed that he was displacing the horse and in many areas of Europe and even Britain he was doing that. But we had forgotten about horses in Hampshire. We wanted something to pull five furrows and implements to match.

Like most firms of this sort, Fergusons used to hold dealer meetings at which we were told about the latest developments and urged on to better efforts. I have never liked the hard sell and used to retaliate to the sermons of the sales manager, one Trevor Knox, with requests for equipment which would satisfy the needs of our bigger customers. Trevor, a fanatical Ferguson man from his early days, was unimpressed. 'Sell, sell, sell,' he used to intone as if he was encouraging the charge of the Light Brigade. Difficulties of the sort I mentioned were, he used to say, a challenge. The Ferguson system would conquer all when properly handled.

In the end I was proved right, but not I think in the way

in which Trevor expected. Suddenly, almost overnight, Harry Ferguson sold out to the arch-enemy of us all, Massey Harris. In this bewildering switch, many of the subordinates who had stuck to Ferguson for years and had followed his principles with a devotion worthy of a better cause, felt they had been betrayed and said so.

The new management – or I should say those of the old management who had remained with the firm – rapidly became the complete Massey men. Like most converts, they would brook no criticism of their new role. There then began a massive redistribution of dealers. In absolute confidence, I was offered the main Massey dealership for the area while at the same time I discovered that the old holders had been offered ours. The eastern end of our area was removed and given to another dealer after we had refused his. None of these offers was for more than a year, after which the dealership would be renegotiated.

We already had a depot at Basingstoke where the other dealer was installed and I thought, sooner than cut each other's throats, we might do better to see if there was some way in which we could combine forces. I wrote to the head of the other firm and suggested a meeting. We had lunch in a local pub where we agreed the stupidity of unrestrained competition; I then suggested that either we should buy their interest out or vice versa. In reply he gave me his balance sheet, which showed a very large sum on deposit. We, on the other hand, were living on an overdraft guaranteed by the directors. On balance I thought that, although we would have a much wider spread of products, sales were much more difficult for us. Farmers had re-equipped since the war and part-exchange, which we never suffered from up to then, was rearing its ugly head. We decided to sell while we had the chance.

The problem was to value the shares. One of my

co-directors was a chartered accountant and we had a London firm as our main auditors. On his advice, we asked our London accountants what they were worth. The Managing Director of this firm came down, solemnly opened an envelope, and said that as far as he could see the shares were worth under £1 – they were £1 shares. I said this was nonsense, the value of the shares surely was how badly the buyer wanted them. So we said goodbye to the accountant, thought of a number, which turned out to be £3, and asked that. In the end we had to settle for £2 17s (£2.85). It just shows how bad the advice of accountants can be. With my share of the proceeds I bought, or anyway paid the deposit on, Hippenscombe farm.

During my time with Anna Valley Tractors, I learnt two management principles which were extremely valuable. The first concerned credit control. I insisted that all accounts must be settled monthly which they never had been. This was because the accounts never got to the farmers until the third week of the month. By that time they had been preceded by all the other accounts and farmers had no money or inclination left to pay. So I said that the month should close on the 26th and the accounts be sent out so that the farmer had them on the 1st of the next month.

According to our accounts department, this was impossible. Every month had to be terminated on the last day. It was the written and unwritten law of all accountancy. My co-directors were equally unhelpful. Being in the motor trade, they were used to customers like army officers who changed posts and disappeared at the end of the month. Farmers, I said, only changed farms at the end of September and seldom then. They were still unimpressed. There was only one thing to do – I told the accounts department that, unless somehow or other the accounts were posted by the 1st of the month, one of them would be sacked and replaced by someone who would do as I directed. The threat worked, and our credit fell from between two and three months' trading to six weeks. When sweet reason fails, there is nothing to beat a bloody big stick. Of course this was before the days of redundancy and all that.

The other thing was servicing. Most modern dealers have very good service departments, but very often we only had the machine in for service when something had gone seriously wrong because of a lack of regular servicing. With my then manager, Phil Long, I concocted a scheme whereby the customer bought with his tractor one or two free services. Then he could buy a further contract on a bi-monthly basis. This worked extremely well. Each of our servicemen, and they were very good ones, was a salesman as well, and the beauty of the scheme was that the customer paid for the serviceman/salesman to visit him. This meant psychologically that he would pay more attention to him than to one who just called for orders.

I was asked to carry on in the tractor business after the sale but declined as I did not like the prospect of working for someone else with whose ideas I was not in sympathy. The management altered as soon as the change was complete, credit control was abandoned, and the service contracts were also phased out. The firm eventually went out of business.

18. Farming Politics

From that time onwards my secondary occupations have been many, but have been fee-earning. They have needed only the investment of my time and no capital. Would my farming have been any more successful had I not become so involved with peripheral things? I really doubt it. I have not, I fear, the quality needed for big business success. I want neither colleagues, partners, nor bosses; I like just as much as I can manage myself. Basically, in spite of all my luck which has been considerable, I am a very timorous operator.

We are always being told that politics is the art of the possible. If that really is the case, the whole of my political life has been dedicated to the impossible. By this I don't mean politics in the sense of national government, but in influencing policies to benefit farming – my farming in particular. Because here I must admit that my one success, and it was a success in straight political terms, was security of tenure. I think it became a success because I was driven by self-interest. I was a tenant farmer, and I was very vulnerable, just as was every single tenant farmer in the country, to the much greater economic power of the moneyed classes. In the end I only bought about 10 per cent of the total acreage I acquired while protected by the Act, the rest came with vacant possession. But I was very lucky to have been in the right place at the right time, when landowners were suddenly seized with a determination to sell.

Many people have reproached me for my activities in this field – particularly when I supported the succession of tenancies recently. The objectors are not tenant farmers, but landlords, their representatives and people who would like to farm but can't. To these I can only say that the whole of life is a struggle that the devil must take the hindmost, and it behoves everyone to look after his or her own interests.

During times of farming prosperity it has always been very difficult indeed for young men without capital to get into farming. That was why so many emigrated as I did in the 1920s and before. Two factors let in hungry young men like myself in the 1930s the slump and death duties. If it hadn't been for these, there would have been no new blood in farming since heaven knows when. No one has a God-given right to farm just because he is rich, is conceived in a landowner's bed, or has been to one of the many colleges or universities. But I can't understand why the same farming and landowning interests which deplore the disruptive effects, as they see them, of capital taxation, fail to see that such taxation, if effective, is bound to create opportunities for entry into farming which any concessions are almost bound to make more difficult. Hypocrisy here is equally divided.

When I arrived as a freshman at Leeds University, the undergraduate scene was dominated by one Jim Turner, a third year agricultural student and a very good boxer and rugger player. Contact with him on the field of play was violent, and after one university trial I decided that if I was going to have to encounter characters like him every Saturday, I had better battle with people my own size in another league, and so I did. Verbal battles are all right, but physical opponents should be picked with caution.

When I entered N.F.U. politics at the end of the war I found the kingpin was a James Turner, who turned out to be the same Jim from whom I had fled the rugger field. Jim Turner, Later Lord Netherthorpe, was President for about 16 years and for much of the time I was in direct opposition to his policies and his style. This was not only in the Council but also outside in the farming press. During all that time he treated me with the greatest fairness, always giving me time to speak and insisting, before an often hostile council, that I had the right to a hearing.

His greatest asset was his skill as a negotiator. At one time I was on the Herbage Seeds Committee and we were involved in endless negotiations with a Ministry which, we felt, was in the pockets of the trade or worse. When things reached an impasse, Jim was called in. He knew nothing about seeds, but in the taxi down to Whitehall we would brief him. Then for the next two hours he would, without repeating himself or calling on us for assistance, present a masterly case, ending with a reasonably satisfactory compromise.

He did this for other special interests as well and I have always thought he would have made a great barrister or trade union official on the lines of Ernest Bevin. Where I was at odds with him was on more general matters of overall policy. He was a great friend of Tom Williams, the Labour Minister of Agriculture, a very nice man personally but one of the most consummate politicians I have ever met. I don't think he ever initiated policy – that was decided by the Civil Service and the Cabinet – but he carried it out.

Farmers are always pessimistic by nature and, as soon as the war was over, began bombarding the N.F.U. with resolutions demanding a long-term policy for farming. Prices were tightly controlled at the time and there was still a fair bit of control of farming through the War Ags. The sum of all this agitation was the Agriculture Act of 1947. The essential phrase in this Act was that the Government would guarantee a fair price for that part of the nation's food which it was deemed essential to grow in this country. The catch here was that the price level was determined by the Government as well as the proportion of food which would be guaranteed.

The Act was Tom Williams' instrument; Turner and the N.F.U. welcomed it unreservedly. However, it did not seem to me that it was any guarantee at all. At the time of its enactment, food was short, and was to remain so for another six or seven years. If there had not been the Act with its price controls and government trading, there would have been a free-for-all during which farmers would have been able to exact any price and conditions they wanted. It was no more than a system of price control and I used to say so, repeatedly, every time the increases awarded at the Annual Price Review looked too small. Farmers used to protest loudly too, but after all the dust and heat Jim would get up and say in so many words that acceptance of the awards would be statesmanlike. I never liked that word.

I wouldn't say that farmers were harshly treated during the period, the crunch came when food became more plentiful in the mid-1950s and the Act had to become a method of price support instead of control. Then Jim found Whitehall a much unkinder place. After a short while, he removed himself to other fields. He set the pattern of N.F.U. relationships with Government for two decades. By his close contacts with Ministers, he established the legend that the N.F.U. was the most effective political lobby in the country. This myth still persists today, but has no foundation in my opinion. What it does do is provide an effective channel for the dissemination of whatever policies a government wishes.

Whether farming lost or gained in the end by these arrangements is impossible to determine, because the alternative was never tried. But I doubt if many farmers were really aware of the thinking behind the food policy which lasted until we joined the E.E.C. in 1973. The aim was simply to encourage sufficient home production to keep the prices of imports well down, while farmers were supported through the deficiency payments to maintain production. From a government's point of view it was, in the circumstances of the time, a success for the nation. I don't really think that farmers suffered all that much either.

The 1947 Agriculture Act effectively removed the possibility of full-scale marketing boards with monopoly powers ever coming into being. Milk was a special case and the Milk Marketing Board was used as a means of securing Government control of the milk market, a task which would have been impossible without it. Again I doubt if farmers really suffered from this policy, but I do not think that they realised how what they thought was their Board was being used.

Although I realised at the time that the marketing issue was lost, I kept on pushing it, both through the N.F.U. and in the Press where I was given plenty of scope. One never knew when, by some aberration, there could be a change in government policy. It was also interesting watching the hierarchy of the N.F.U. wriggling off the hook of commitments thrust upon them by the members at the Annual General Meetings and in Council. They (the hierarchy) realised, but refused to admit, their impotence.

One result of the frustration of the time was the formation of the Fatstock Marketing Corporation, now known as F.M.C. This was launched with a great fanfare as being the answer to the challenge of the times. As soon as anyone talks about something being a challenge, you can rest assured it is going to be a dead duck. It began inauspiciously because the founders overestimated both farmers' loyalty to any marketing system which was not compulsory, and the ability of a meat wholesaler to control a free market unless it had monopoly powers.

I made myself very unpopular at the time by pointing out that no wholesaler could pay more than the market rate and that farmers would, as they always had done, give no more loyalty to F.M.C. than the payment of an extra penny a pound could command. I approved of its formation because as a fairly large-scale meat producer I was pleased to see yet another wholesaler competing for supplies. To that end I subscribed some capital and am still a shareholder. I refused to sell my shares to Borthwicks recently, whose tactics were obviously to reduce the number of buyers. As a wholesaler I think F.M.C. has done the task for which it was suited if not designed. However, it has been a disaster as a bacon curer. This is because the management, under the late Sir John Stratton, decided to defend the Wiltshire cure to the last drop of producers' blood, just as Walls under Lord Trenchard decided to make a principle of the heavy cutting pig. Maintaining principles regardless is a costly business in any sphere.

A few years ago, in company with some Hampshire farmers, I was concerned with setting up a rape seed growing and selling co-operative. It was quite successful, but only I believe because we refused to take anyone as growers whom the individual directors did not know to be reliable. They had to be prepared to deliver to us, even if the market price had meanwhile gone above the contract agreed. Other groups exist for peas, grain and so on. If they can be kept small, as the rape growing organisation was, they work. On a bigger scale, they haven't yet.

Incidentally we nearly came unstuck in our first year. We had contracted our expected output to a London broker at a fixed price. One day I noticed in *The Financial Times* that this firm was being sued in some United States court for a fantastic sum. We made enquiries and found that this was indeed the case. They had been drawn into the ramifications of a massive scandal by no fault of their own, but if the case was lost they would be in deep trouble. This made things very awkward for us as we were contract bound to deliver, even if the case was not yet finished.

To make matters even worse the market had dropped after we made the sale and, even if we had delivered elsewhere, we could not have paid the price which we had promised our members. Our mistake had been not to have insured payment of the contract at the moment we made it. After a very nasty time, which the Committee kept to themselves, we managed to insure the contract through a friendly merchant bank for very little money, and that was the practice ever after. It just shows that co-operative trading isn't all that easy.

In some countries – notably Denmark and Ireland – the co-operative principle is almost a religion. In others, it is especially favoured by Governments so that it can not only demand, but reward, farmers' loyalty. But British farmers have always been comparatively well off and thus can afford the luxury of individual trading. The British road to co-operation is littered with costly failures, mainly because those starting them failed to appreciate this essential truth. It is possible in time that smaller specialist Co-ops, such as that for rape seed, will succeed and gradually become bigger and bigger.

I gave up active participation in the N.F.U. in the mid-1950s, preferring to stir the pot from the sidelines. To some extent this is a confession of failure. I was, I was often told, abusing men who were spending the best years of their lives looking after farmers' interests – mine was an irresponsible attitude. I am sure that I should have stayed dutifully on the Council doing all the donkey work – negotiating the nitty-gritty of such things as animal health, seed standards, wage agreements and the like which form 90 per cent of union work – but I didn't. I honour those who do, reserving the right to criticise them. Actually, this is not quite true. I always criticise policies which to me appear misguided, but never directly those responsible.

In many ways I think that a continuing close relationship between the Minister of Agriculture and the President of the N.F.U. has its dangers for the industry because policies – once agreed – may not be discarded just because it would be a pity to cause an upset. Where even more care is needed, is in relations with commercial interests.

While I was on the Council, it came to the notice of the Commercial Committee of the Union that some fertilisers were much cheaper than others in terms of unit costs. The information was there on the published analysis for those who had the knowledge to work it out; but few farmers had this at their finger-tips. The Commercial Committee tried to get an article on this subject published in the N.F.U. paper but this was squashed by Council on the grounds that it would do public relations no good. We were also told that we might lose some advertising from the interests concerned. This clearly was not good enough. What was the point of having an N.F.U. publication if it could not give members information of this sort? The matter was raised at Council again: this time the Council reversed its decision and left it to the President to sort out the matter.

Sorting it out took the form of a private lunch at the Savoy with some officials of the firm concerned. There our hearts were wrung with the accounts of the difficulties faced by some firms, that without them there would not be enough fertilisers to go round, but that eventually they would have the means to get even with the manufacturers who were producing cheaper material. My colleagues and I were unimpressed. I was their spokesman and insisted that our only duty was to look after our members' interests even if they could not look after themselves. If we failed to publish for whatever reason, we would be failing in our manifest duty. At this point the Managing Director of the firm concerned walked out, and we won the day. A lot of nonsense used to be talked about the identity of interests between farmer and suppliers and customers. So there is. But the prime reason for the existence of these interests is to make money out of us.

In some respects, the N.F.U.'s concentration on the political scene has replaced consideration of the changes in the commercial world. The concentration of the buying interests into a few hands of overall ownership could, in the end, lead to a lack of competition both in markets and supplies. More interest should be taken in the relative advantages of the multiplicity of chemicals which we shower on the crops. The present Seeds Regulations, imposed by *Diktat* from Brussels under E.E.C. rules – regulations which are largely disregarded on the continent – are a sign that the N.F.U. is getting some of its priorities wrong.

There is no reason why this should be so. While contact with a British Minister of Agriculture is essential, it must be realised that neither he nor his Government is a free agent. The fact that a British Government can control the rate of devaluation of the Green Pound is a temporary phase. Overall the future of British farming policy is going to be decided by the Council of Ministers of nine or even 12 countries of very divergent interests. In Council, the British Minister represents no more than about 3 per cent of the total number of farmers in the Community and will have to strike bargains which take that fact into account.

The N.F.U. will be needed more than ever in the future. It deserves the support and the money of all farmers, because it seems to me that, politically, it is very weak both in national and European terms.

19. I Break into Journalism

One of the problems of running a fairly large farming business – and by the mid-1950s and 2,000 acres mine was large by most standards – is finding something for the boss to do. Management is simply a matter of finding the right men and leaving them alone. Too much management brings out ideas, and ideas cost money, so by the time my farming had reached that state, I was often more of an encumbrance than anything else. This mortified me no end, because I liked to boast that I could do any job on the farm.

This is strictly true. But I am not very handy with my skills. My ploughing used to weave all over the fields and I was never very good even at lapping my work with harrows. At one time I used to keep the combines running through the lunch hour. On the last occasion I did this, we were thrashing rye grass seed. The combine was fitted with a pick-up in front and a side-rake to get rid of the straw behind. The field was fairly sloping and, as I was coming down one side, the combine began to run away.

This particular model had a hydraulic speed control and a gadget to raise and lower the platform, side by side on a rail in front of me. I pulled the lever and immediately, instead of slowing the combine, the platform lifted and the machinery gathered speed. I pushed it back and grabbed at the other. Unfortunately I pulled it over centre and, instead of slowing, it accelerated. The only thing to do was to run across the crop to reduce speed. The ensuing mess blocked the combine drum solid, and churned up a good deal of the crop. By the time the men came back from lunch, I was still trying to clear the drum.

I apologised to the driver who gave me the sort of pitying look that the experts give the useless. Various remarks were passed about the difficulties of keeping up to schedule with amateurs on the driver's seat, at the end of which I announced that I would never drive a combine again, and that between them they would have to find spare drivers when necessary.

Since then I have only joined in when absolutely necessary, I can run the drier, particularly at night, and that is about as far as I go in that direction. I lamb a ewe occasionally and, if the pigman is away, do the routine injections without which no farm animal can possibly survive these days. Otherwise my duties are entirely bureaucratic: the accounts, the correspondence and marketing. Even this does not take long: my pigs are sold by contract and renewal is a matter of a telephone call. Sheep usually go to the same outlet, although I do check the prices offered in other directions. Grain is all sold by Christmas if not before. So what am I to do?

It is this lack of practical work, of achieving something with your hands, that drives so many farmers to N.F.U. and Local Authority work; or to hunting, shooting, fishing and other not such innocent pastimes. Prosperity in farming circles has probably caused more divorces than any other single factor. If you are a 100 acre farmer milking your own cows and sharing the load with your wife, neither has the time or the energy at the end of the day to look over the fence.

Harvest in 1955 was easy and by early September I was sitting in my office trying to think of a good excuse as to why I should not take the children to the sea for a few days. Then the telephone rang. It was the news editor of the *Farmers' Weekly*, Geoffrey Eley, asking me if I would like to go to Blenheim and do a write-up of the Duke of

Marlborough's new milk bar. As it happened I could not do that job, but I said I would like to join a party which was going to Russia. That he said would be a good idea. They had no one available and had thought of asking me to go, but were sure I would be too busy to go off for two or three weeks.

I had known about this semi-official party and had tried to get invited on all sorts of pretexts, but was already too unpopular in establishment circles to be considered. I agreed to go with a photographer and then told Ronie that instead of going to Swanage I was going to Russia.

Going to Russia is quite old hat these days, but at that time it was something of an adventure; nor was it easy for journalists. The photographer, Gordon Craddock, and I presented ourselves at the Soviet Embassy in Kensington Gardens only to be told that a visa would take at least three weeks. As the party was leaving in three days, we more or less lost interest. However, after the rest of the party had gone we were told we could start three days later.

We flew as far as Helsinki where we met another barrier – no seats on the flight to Moscow for at least three weeks. Three seemed to be a magic number to the Soviets. Then I remembered a story I had read about a train to Moscow. We went to the Finnish travel agency, sat down and refused to leave until we had tickets. Sure enough the next day we joined the train and spent the next 36 hours slowly bumping along to Moscow.

You are never free of Big Brother in Russia. An Intourist guide picked us up at Leningrad and saw to it that we had a hotel room at the National in Moscow. The National was occupied by frustrated journalists trying to get either further into, or out of, Russia. Once installed I tried to locate the rest of the party; the obvious place to try was the British Embassy, so I telephoned. A cultured voice answered my query with the information that they had all gone to the Far East that morning and that we were to remain in Moscow. When I hung up, I turned round and saw one of the party we were trying to join standing behind me at the Intourist desk. He wasn't in the Far East, but he was reliably informed that they were all going to Leningrad that night. How to join them?

The administrative office for the party was in room 903 at the Moscow Hotel across Red Square. When I arrived there, the only occupants were three solemn Russians. I tried my best English, French, German and Spanish to no avail. So we just sat and looked at each other. After about half an hour a young woman came in and asked in very good English what the matter was.

I explained our predicament and that we were representing important British papers. I mentioned *The Times* for whom I sometimes wrote, in fact I emphasised it. The boss-man shook his head.

'He is sorry,' she said, 'but you could have come to Leningrad tonight – only your passports which had been taken at the National Hotel will not be back for three days.'

Then she said she was very sorry but she had to go. I grabbed her arm, returned to the attack, and explained how the Great British public was dying for an authentic account of life in Russian farming, how great sums of money had been invested in our journey, and we had not come to Moscow just to watch the Dynamos.

I must have touched dear Luba's heart because she then set into those three men with a vigour which would make me hesitate before ever getting involved with a Russian woman, however beautiful. After about 20 minutes, while the three men looked more and more uncomfortable, one of them said something to her.

'He asks,' she said, 'if you have any British coins or stamps.'

As it happened we had, and emptied our pockets from which he took a sample of each. In half an hour we had been accepted, our passports were back and we met the rest of the party. I lifted a pair of woollen gloves from John Mackie which he had brought to give to the interpreters – Russian women were apparently starved of such luxuries – gave them to Luba, and we were on our way.

I wrote several articles for the *Farmers' Weekly* as a result of the trip and found the whole experience most exhilarating. It is one thing to travel as so many do, but to travel with such a purpose and to be paid to do it, is to my mind the height of human enjoyment. The beauty of journalism is that, in most cases, people are pleased to see you. If you represent the B.B.C. or anyone else, it will open doors which are closed to the run-of-the-mill tourist.

The experience stimulated me to make a real effort to break into journalism on a consistent, and not a spasmodic, basis. I realised early on that, if I wanted to enjoy the high spots such as the trip to Russia which had come out of the blue, I would have to earn the regard of editors. Then I would be considered when and if such opportunities arose. This meant accepting any jobs offered, however dull. I must have sat through dozens of conferences and other functions, and I sent in some pretty tedious reports. I also covered shows both here and abroad for the *Farmer and Stockbreeder*, particularly the machinery sections. It was occasionally boring and of course very underpaid, but it kept me occupied doing something practical, and learning such skills as journalists need.

Then I had a stroke of luck. When the B.B.C. began Farming Television, I opened the series and immediately the publicity put me right in the farming public eye. At the time I was writing regularly for the *Farmers' Weekly* once more, but the Features Editor cut the price of one of my regular articles, because he said he had had to use a photograph on the page. When I complained, he said that, as I was a surtax payer, I had nothing to worry about. I was furious, and while I was wondering what to do in retaliation, the *Farmer and Stockbreeder* offered me three times the F.W. rate for a fortnightly article and I have written for them ever since.

One of the essentials in writing for farmers is to know the subject as a farmer and to write in farmers' language. Some farming writers are very good indeed. But I always think myself that articles by people who are risking their own money in the job carry far more conviction than those of people reporting the activities of others. This goes against all the convictions of the National Union of Journalists, but it is a fact.

The B.B.C. TV farming programme has always been chaired by farmers and the interviewing is by farmers too. All of us who do it have large farming businesses and presumably don't need the money, which in any case is not very great. What I am sure we need is occupation, and I never grudged the 40 to 50 days a year which I have devoted to B.B.C. work for the best part of 20 years. I don't think the farm suffered, and the editors and producers obviously appreciated my efforts and paid me for them. This made me feel wanted, which I certainly wasn't on the farm.

But it was far from plain sailing. I went to Australia and New Zealand, sponsored in part by the B.B.C. and the *Farmer and Stockbreeder*, in late 1957 and, while driving through the Manawatu Gorge in New Zealand, felt most dreadfully ill. I recovered after a while but was struck with the same pain and nausea in Pitt Street in Sydney and went into a cinema to recover. None the worse the next day I travelled home and carried on as usual. Then, while filming a programme in Birmingham, I was struck again. I recovered somewhat and drove home. The next day I went to hospital with a burst appendix and did not surface again for nearly a year. Radio and television had to stop but, in lucid intervals, I continued writing. Being as seriously ill as I was, made me appreciate the unselfish devotion of the doctors and nurses who saved my life and to them I am forever grateful.

The *Farmer and Stockbreeder* was going through a difficult time and, while television work continued, I felt I wasn't getting on very fast. This was because I was unfortunately typed as a farmer. I knew I was a good interviewer by this time, I was well up to date in current affairs, I even knew a lot of what is loosely termed political economy, which is how the world works. But no one would employ me on any other subject than farming. Like the clown who traditionally always wants to play Hamlet, I wanted to be a general interviewer, to travel and take an interest in other things in life than crops and food. It was no good: I even appeared quite a number of times on 'Any Questions' but somehow I could never answer the questions in a farmer's way, or with a Hampshire or Dorset accent. So I never became a regular member of the team.

Eventually I had two strokes of luck. The B.B.C. decided to make a film about the Common Market, set in all the member countries and Denmark, and *The Times*, or rather its retiring Agricultural Correspondent Anthony Hurd, asked me to start contributing. I wrote several articles for *The Times* and found the work interesting because I was writing for the non-farming reader and the presentation had to be different. The reader, Hurd told me was probably a senior Civil Servant travelling to Brighton looking out of the window and wondering what was going on on the farms he could see. Nothing must be left to the reader's own grasp of the subject. This sounds elementary, but it is surprising how few journalists can interpret subjects so that the layman can understand them.

The B.B.C. job was fascinating. I spent nine weeks travelling in Europe with the producer, Ronald Webster, picking up crews in each of the countries we visited. I always thought Ronald Webster a most talented producer and he taught me a great deal about TV reporting. Unfortunately we were not too compatible when not working, and from time to time relations were strained – probably because we had a small breakfast, filmed all day, then ate a huge meal and drank a lot at night.

The product, six half-hour films, was a success and was sold round the world. It was probably the first picture of modern European farming. It also set me up as a Common Market expert. I was called upon to write or speak whenever the Community, of which we were at the time an applicant, was in the news agriculturally. Regular journalists without my background knowledge must have been just as frustrated as I used to be. At the height of this success, I tried through my acquaintance with various producers to join the 'Panorama' team of reporters.

This was squashed by the hierarchy at Birmingham who stated, in no uncertain terms, that my fees equalled those paid elsewhere, and that I should be content doing this. However, I was promised a programme at peak time on B.B.C. I and assembled a forty-five minute film out of the films we had made. Before it could be shown though, de Gaulle turned us down, and the film was withdrawn.

During this filming was the only time I found any evidence of B.B.C. censorship. While we were abroad we had a day off, in Stuttgart, and I wrote a letter to *The Times* pointing out that, in the course of many weeks travel among European farmers, I had found no evidence of any enthusiasm for European unity. It appeared to me that everyone was looking to see what advantages they could get for themselves from membership.

There was at that time a lot of correspondence on the

benefits of membership and this letter caused a certain amount of decorous commotion among pro-Europeans. It also attracted the attention of the B.B.C. Eventually I was called to account in a meeting over a lunch in Birmingham. I was told that, as I was under contract to the Corporation at that time, I should not have taken part in any controversial matters – particularly one on which I was actually engaged in working on the B.B.C.'s behalf. This I thought was utter nonsense and said so. This was not accepted. I should read the small print at the back of the contract. I did, and it only said that I should not become a candidate or take part in national politics during its currency. Then I suggested to the Controller, John Dunkerly, that he should write me a letter stating that, while working for the B.B.C., I should not have any ideas of my own on matters of national interest. This he refused to do and in the end the affair, as usually happens, fizzled out.

Although the B.B.C. lost interest in the Community after the de Gaulle turn-down, I kept up my studies. I visited Brussels once or twice and got to know some of the Commission officials which was helpful when I joined *The Financial Times* in 1962. This came about quite by accident. I had been driven off the Test by a thunderstorm and was offered shelter in a Jaguar belonging to another angler. He produced some gin and we started talking about this and that, and eventually focused on farming and the Common Market. After a couple of hours he asked me if I ever wrote, I told him I was committed to writing regularly and to broadcasting for farmers. He said nothing further and I thought no more about it.

Three weeks later we met again in the same place, and he asked me if I would consider writing for *The Financial Times*, of which he was Editor. The proposition was attractive, a deal was made and I have been with them ever since. Gordon Newton became a close friend, but this friendship was never allowed to interfere with my business with *The Financial Times*. I was given certain guidelines to work to and as long as I didn't try to grind any of my axes too obviously, I had a completely free hand.

He taught me one supreme lesson. F.T. writers are encouraged to use the library for statistics and evidence to back their assertions. As few of the readers have these facilities, there is little fear of contradiction, which is probably why the letter columns are so dull. One day in 1969 there was no butter surplus in the Community, and I wrote a feature, well garnished with statistics, graphs and all the rest, which demonstrated that the butter surpluses would never happen again. Gordon read it and told me it was nonsense. 'But why?' I asked; 'Here are all the statistics of falling cow numbers and people leaving the land.'

'I know you have done your homework,' he replied 'but your final assessment is still nonsense.' 'Please explain it then.' I asked.

'Years ago,' he told me, 'I was Commodities Editor and employed half a dozen specialists, as you are, to write about specific subjects. They were always wrong although they knew everything. I only knew one thing but was always right.' 'And what was that?' I asked.

'Commodities go up and down, it's really all you need to know.'

It was a big and prosperous paper at that time. This meant that it devoured editorial matter like a shoal of piranhas are said to devour human beings. If I was willing to write anything, it was always published and paid for. I discovered that the favourite contributor for any editor is the one who will fill a regular spot without fail. It is one less worry on his mind.

I was still, alas, confined to writing about agricultural matters. The F.T. employs some notable economists. They have one thing in common, which is that they have never, as far as I know, been concerned with running a business on their own as I was doing. I would love to write on the best methods of getting Britain back to work, to stop strikes, to solve the balance of payments, and anything else one can think of. But, even though I am one of the oldest members of the staff, I am still confined to farming.

I doubt if my writing has altered the course of history one millimetre, but it has afforded me the greatest interest. I have travelled extensively, reporting again on the food and agricultural scene and I am, thanks to the F.T.'s world-wide circulation, one of the best-known and (I hope) best-informed agricultural writers. I have been very lucky. So incidentally have my staff and family – if I hadn't been so occupied I could have been an awful nuisance.

20. Where is Farming Going?

I find myself in a dilemma over the Common Market. Membership, even with prices reduced by the Green Pound, has made my own farming very much more profitable than it would have been if subject to world prices and deficiency payments. But matters of national importance should be judged on an overall, not a sectional basis. The balance of interest has been loaded against Britain in a way which certainly wasn't foreseen by those who advocated entry at almost any price.

On the face of it it must be wrong for a world trading country like Britain, which has to import 40 per cent of its food, to shut itself deliberately off from its major suppliers and bind itself to the highest price food sources in the world. When you point this out to the Eurofanatics, they claim in effect that membership has removed the threat of European war.

This is dangerous nonsense. The major unifying force since the end of the last war has been the threat of Russia and the atomic bomb. Unless NATO is made to provide an effective defence against Russia, the Common Market can do nothing for security. In fact many of the members of NATO are not in the Community and France is not a member of NATO.

Well then, say the 'pros', membership will destroy the evil of nationalism, the tribal instinct. Don't you believe it! Looking at the countries which make up the Community today, there is plenty of evidence that nationalism is far from dead. Even in Britain, after centuries of union, the Scots and the Welsh, or a large proportion of them, are seeking means of separating themselves from the centre and doing more and more of their own thing. The only country in the world which I know where nationalism – or should one say the tribal instinct has been suppressed – is the United States, where the various ethnic groups lost their identity through emigration, and a vicious civil war unified the country.

Taking the farming angle for a moment, the United States is an interesting example of a Common Market. On one of my visits there I travelled north from New York into New England. Much of it is poor hilly land covered with secondary scrub. If you go for a walk through these woods, you constantly come across evidence of earlier farming: houses, barns and walls. It is the same in the State of Georgia where pine-trees are taking over the contoured fields.

Where has all the farming gone? To the good lands of the Middle West and the irrigated deserts. Few American farmers work land as poor as mine, so full of stones. In fact the sight of what they call 'my rocks' brings tears to the eyes of my American visitors. This has been due to common prices throughout the U.S.A.

If you transpose the American situation to Europe, you would find that much of the poorer and more difficult land is situated in Britain. Climatically and physically this is a poor country and likely to remain so. When, if ever, common prices are fixed right through the Community, and are fixed by economic and political criteria, much of Britain, particularly northern and western Britain will be likely to go out of production. This may take a long time to come about, but it is an almost certain eventuality.

Then no one can say that the benefits of scale of size in any organisation either commercial or national are particularly obvious. To increase the size of anything means an increase in administration, and this means more bureaucracy. I have nothing against bureaucrats as such.

At home they may be the nicest or kindest of men. But assemble them in offices and give them power and they illustrate all the finesse of a herd of elephants. Their power becomes immense, they are safe in their jobs, unlike politicians, and they go on until they die or retire to be replaced by others trained in the same mould. The British Civil Service, which in some sections has probably done more harm to the country than any single outside enemy by bad advice given to Ministers, is now effectively, through Community membership, subordinate in many respects to the Commission and probably will be to the European parliament when that body comes into being. It is difficult enough now to move the British bureaucracy into positive action. Once it can pass the buck to Brussels or Strasburg, it will be impossible.

What I should have liked to see as a foundation for European unity would be some form of customs union, inside which each country could do its own thing and from which eventually a closer union would have evolved as European countries became educated to it. Attempting union by decree as the plethora of regulations emanating from Brussels is attempting, is setting the scene for an eventual revolt.

My own studies of the Community have been to do with the Common Agricultural Policy. This was originally set up with two main objects: to improve the living standards of the European farming population, and to provide for the unity of Europe. The first of these objects has been achieved and was reached quite soon, not because of the C.A.P. but because the industrial boom, which ended in the Arab oil crisis, drew about half the farmers of Europe into industry.

Since 1974 the farming populations have remained more or less static, mainly because member countries have seen to it that no more farmers than are necessary leave the land to add to the ranks of the urban unemployed.

They have done this by insisting on much higher prices than are necessary for efficient food production, and special national measures which effectively defeat the Community's rules on competition. The most obvious of these is the advantage taken of currency instability and the manipulation by each country of the consequent Green currencies.

The original instigator of the C.A.P, and its first Commissioner was Sicco Mansholt. Once an attaché in London, later Dutch Minister of Agriculture, he literally towered above any Commissioner since. I interviewed him a number of times. On the first occasion in a television studio in Brussels, he almost had me in a spin. He has a trick under questioning of bulging and then retracting his eyes. Spellbound I watched him, and suddenly I thought of John Buchan's book, *The Thirty-Nine Steps*, where the villain was able to do something of the kind. Luckily I had the question line on a card on my knee and was able to carry on. Mansholt asked me afterwards why I had wavered slightly as he put it, and I said I was just altering a question to fit his previous answer. Not that I thought him the villain of the piece!

He was an impressive man by any standard, of immense dignity. And he pulled no punches. If he addressed a meeting he simply dominated it. Early on, he saw the danger of creating food surpluses and said that the way to prevent food production increases was not to reduce prices, which in general stimulates more production, but to remove land from farming altogether. His famous plan, which laid down that some 12,000,000 acres should be taken out of production throughout Europe, was the logical answer to the situation as it then was.

He was defeated by the mass political objections of Ministers of Agriculture and by a sudden shortage of butter which for the only time in the E.E.C.'s history kept it out of intervention for a few months. He left the

Commission, after being its President, soon after British entry; and caused a lot of embarrassment at a celebratory dinner at Hampton Court by claiming that the Community was developing into a self-centred club of the rich. Since then he has retired and has refused to give any interviews.

His immediate successor was an Italian; then came Pierre Lardinois, another Dutchman, once again a diplomat in London and the Dutch Minister of Agriculture. Lardinois was a great pragmatist, largely devoted to looking after farming, particularly Dutch farming, interests and without any deeply held convictions about Europe that could be discerned.

His successor at the time of writing, Finn Gundelach, is one of those infuriating bureaucrats. He was a Danish Civil Servant, who seems to be completely devoid of personality. He clings to his principles with a perseverance which can only be explained by his complete identification with the Commission, and loyalty through thick and thin to the rules of the C.A.P. Because it absorbs at least 75 per cent of the Community budget, the C.A.P. must not be tampered with on any account, according to him and his agricultural officials. They claim that without it European Unity would have lost its purpose. What they mean is that the Agricultural Commissioner would lose a lot of his importance and the Civil Servants who administer it could well lose their jobs which include probably the best tax-free salaries in Europe.

For instance, the only sensible solution of the dairy problem would be to impose some sort of national quotas on production so that individual countries which produced too much milk would suffer a reduction in price. The difference between the Community milk price and that in the rest of the world is a matter of at least 100 per cent; so that marginal production over requirements could incur a very heavy penalty, such as paying only world prices for the surplus. If this was spread over Community production as a whole, the fall in every farmer's returns would be minimal. But if it was confined to the offending country, or even to the offending farmer, the effect might be to cause some fail in output. It is generally believed that the officials of the Commission see this as the only answer, but daren't admit it publicly. In any case, according to Gundelach, it would defeat the Community's rules on competition. But then the rules of competition are continually flouted by the member countries in the most blatant manner.

None of them is worse than Germany. In late 1978 I was a guest, with other journalists, of Josef Ertl, the country's long-serving Minister of Agriculture. The real purpose of the exercise was unclear, but there is no doubt that, as an exposition as to how Germany looks after its farmers, it was first-class. By British standards of low-cost farming, Germany is a most inefficient producer, the tiny farms and herds are very high cost indeed. German farm prices are 50 per cent above the British and 11 per cent over those of the next highest priced Community country – then Benelux.

This wouldn't matter if British and some other countries' taxpayers were not having increasingly to finance the cost of Germany's growing butter and beef mountains. It is not all West Germany's production either. For some reason the Iron Curtain, while preventing human access to the West, does nothing to interfere with the flow of trade between the two Germanies. Surpluses from the Communist bloc, even as far away as Rumania, are sent through East Germany and change their ideology on entering the Community's cold stores without levy or other impediment. When I asked Ertl the extent of the trade and its cost to Community funds, he said that no records were kept.

But don't fall for the myth that European farming is based on inefficient peasants milking two cows on the slopes of the Massif Central; a myth which is continually repeated. In terms of economic efficiency, that is production per hectare and per beast, German and most other agricultural systems are fully efficient and a very serious competitive threat to ourselves. British agricultural efficiency has largely been concerned with livestock farming – in particular pigs and poultry. Our units are the largest in Europe; the Eastwood (now Imperial) and other empires are the biggest in the Community and probably the world in terms of individual enterprises. These empires were built up on cheap imported grain and other feeding stuffs, which have been available more cheaply here than in most other member countries except for Holland.

European grain prices have always been much higher than those on the world markets and their livestock industries have had to rely on other feeds and forages. Except for the Dutch who, with the uncanny shrewdness of their race, arranged for a supply of manioc and other tropical carbohydrates which could replace cereals and would not carry the full cereal import levies. It is a fact that, in Europe, the production of every unit of meat, milk and poultry uses less compound feed than does Britain today. In other words the feeds are home-grown. British producers are not yet meeting the full Community cost of animal feed.

When they do, they will come up against another serious hurdle, particularly if competition comes into it. At present fresh pigmeat and some other animal products are kept out of the U.K. by disease regulations. In terms of production per unit of feed, there is probably little to favour either country, but in labour costs the advantage is all with the Europeans.

That is because no labour is employed to speak of, especially in the livestock sector, because production is by the family farm. And a family, which has the intention of staying in farming – which most of them have – is almost impossible to get rid of by the ordinary exercise of economic forces. If, in addition, the Governments of the countries concerned, are, like the Germans, determined to keep a sizable population on the land; there is just nothing that can be done about it.

So far the pro-Europeans have refused to face these facts and keep blathering on about competitive efficiency as they doubtless will do as the bailiffs are knocking at the door. Some will point with pride at the arable sector. Our arable farmers are very good; but they too employ labour and, in the European context, don't have very good yields. In sugar-beet and potatoes, we are well down in the league; in cereals, just about average. This isn't our fault, we haven't enough good land.

In fact the last thing we need is free competition with Europe because then many of our most efficient livestock enterprises will go to the wall and so might some of our arable farming on the thinner lands. I believe the only sensible way in which to get some form of Common Agricultural Policy working is to revert to a customs union with some national quotas or market shares reserved and the responsibilities firmly shouldered by the National Governments concerned.

As a last argument the 'pros', while admitting the logic of my thesis, claim that there is no need to worry. It will all be taken care of by the growth of population, by the demands for a better standard of living, and the consequent improved market for all food stuffs. I think this to be nonsense, as I shall show in the next chapter. Food production has always caught up with effective demand, that is demand which can pay for it.

There is a final point to be made. By joining the Community, this Government effectively handed over large sections of policy to a Council of Ministers meeting in Brussels. Without becoming emotive about it, this is the first time this has happened since the Norman Conquest. However well meaning British Agricultural Ministers are, in the future they will only be able to look after British interests to the extent allowed by their fellow Ministers in Council.

21. Lessons from Abroad

One of the great pleasures and tangible benefits of journalism has been the opportunity of making friends and meeting farmers in other countries. The first of my overseas ventures after the war was not, in fact, journalistic. The N.F.U. was host in 1946 to a number of overseas farming organisations at a conference in London. From this was formed the International Federation of Agricultural Producers: I.F.A.P. for short. I doubt very much if I.F.A.P., for all the brave talk of co-operation at its foundation, has ever done much for farmers here or anywhere else. The organisations represented have too many conflicting interests to be capable of any concerted action beyond paying lip-service to ideals of no possible attainment.

Still it is something that representatives of American, E.E.C., Australiasian and other farmers should meet every now and then to wine and dine each other. It does at least widen their individual horizons, and I think that the N.F.U.'s subscription is probably justified for that reason alone.

At this first meeting I was filing into a lunch which was of typical post-war austerity; a little Spam, a lettuce leaf and potato; and heard some French people in front of me discussing the food with all the scorn that they reserve for English attempts at what they have always considered to be their own speciality. One of these was a delegate who had charge of their interpreting. I joined in the conversation and introduced myself.

She was Anne Bouchardy, a Belgian married to a

French landowner and a very strong personality in French farming politics of those days. During the war, while her husband François had been occupied with a family business in Turkey, she had run the estate. She had also formed part of an escape-route for Allied airmen and others. She had been decorated both by the French and Americans for her services, and was a very considerable person indeed. I met her again the following year, at yet another conference in Holland, and later made the first of many visits to her château by the Loire.

François became a close friend as well, and I spent a lot of my time during my visits driving around the estate in a sort of two-wheeled trap with a hood seeing how things were going. He was a practical farmer of a rather old school originating from the Jura. His philosophy was a simple one. The only thing to invest money in, he used to say, is land or livestock. These can grow into money, while machinery – to which the French farmers had taken enthusiastically after the war – will only rust away.

These sentiments were the despair of the farm foreman who was for ever asking for a new tractor. But François would point with glee at a horse he had bought ten years ago which was worth twice its original price for meat.

'Show me the tractor which appreciates over ten years, and I will buy one Paul *mon garcon*' he would say.

Then he made a visit to Turkey and, much to his fury Anne mechanised the farm to some extent in his absence.

Both he and Anne have gone now, but his remarks made a great impression on me, or rather they reinforced my own basic inclinations although, for much of my life, I had had so little capital available that the choice was not really mine. But the theory behind it is right. Land and livestock – and by land I include fertiliser and other crop inputs – are the basis of production. Machinery and buildings most definitely are not.

The only time I bought new machinery in any quantity

was to take advantage of the 100 per cent depreciation allowance. This allowance is a delusion after all. After you have bought a machine, you have only saved the tax on the amount you have spent, say 40 or 50 per cent. You still have to find the balance. In many cases I remember, it seemed better to avoid buying, to pay the tax and keep the balance yourself. Most machinery made in the postwar period is good and reliable; it will last many more years than the average farmer keeps it. I noticed particularly in Australia, Canada and even the U.S.A. that farmers had much older machines than mine. Twenty and even thirty-year-old combines were still at work and likely to be so for a long time yet.

There is no sense at all in having a big investment in new machines. They depreciate very fast and, while you can always raise a loan on land or liquidate cattle or other stock, there is nothing worse to cash than machinery on a bad trade. I know that it is considered best in an inflationary period to keep investing in new machines, because if you don't in the end you may be left with a lot of old junk and no money to buy new ones with. This is certainly true, but I often wonder if owning machinery is sensible at all.

Years ago I noticed that the contractors who built our local by-pass and those who worked on motorways hired their machines. In fact the plant hire industry is a very large one. So in 1965 while driving into town one day, I suddenly thought the best thing to do would be to hire a couple of new combines rather than buy them. I approached the local agent and offered him £1,000 a piece to hire two new Ransome combines for the season. The basis of my offer was not a result of long calculation as to the relative merits of hiring or owning. I thought that for those days with about 800 acres of harvest 50s (£.50) an acre was a fair price for the cost of combining. I also thought that this figure would tempt him to make a deal, which it did. I carried on on this basis for several years and then bought the last two Ransome combines made because he refused to play any more. I also hired sundry tractors and other implements. I used to get the best deal I could by offering to pay the hire charge in advance. It wasn't that I was short of money, sometimes in fact I had a credit balance, but I wanted to keep my capital for more important things, such as buying more sheep and pigs.

Another thing I noted on my trips to Australia and Canada was that few farmers stored their grain at home unless they had to. On the other hand, I could store the whole of my harvest and still can. I used to pride myself on the fact that I could keep it sweet until the next summer. This was one of the most stupid policies that I ever adopted. The premiums paid for storage, even under the old Home Grown Cereals scheme, were nothing like enough to pay the interest on the money value of the grain, let alone cover the physical costs or the wastage. The bank, of course, was quite happy to have a well-secured loan which could be quickly liquidated and the users, millers and compounders were highly delighted. Very occasionally I could make a small killing through some temporary shortage, but generally speaking it was a dead loss.

So I made it a rule that I would sell everything by the end of the year and make sure the money came in by the end of January. The results were surprising. I had a credit balance and the cash flow, although I did not call it that, looked surprisingly well. So well in fact that, over three years, I built four new cottages for my men, a new grain storage barn as well, and never increased my overdraft at all.

This obsession with the weight of an overdraft is simply a reflection of my inherently peasant mind. Some years ago a very good accountant told me that, in my position and at my age, I should be well advised to have a very large overdraft and that if I died with one of £100,000, my heirs would all be very much better off. I thought about this for a while and then decided that, when an expert gives you advice, the best thing to do is the exact opposite. In the event interest rates escalated and, had I had the overdraft he mentioned, I should have had to work quite hard to pay the interest, which would have added £15 an acre to the rent of my present farm.

Don't assume from what I have said that I distrust all the lawyers and accountants who spend so much of their time advising us how to avoid losing so much of our wealth to the Chancellor. These men are simply supplying a demand to the best of their ability. I often think their abilities must be limited because, if they were as clever as they are claimed to be, we would be working for them and not the other way round. In any case, they can only set out the choices. It is up to us to take our pick.

I also have my doubts about the morality, although this is hardly the word of avoiding too much tax. Not because I think the country deserves the revenue – the politicians would waste it in any case – but because the capital taxes are a form of redistribution. I have always been conscious of the fact that my good fortune in being able to build up a farming empire was due to the Depression and to death duties. Those who claim that these taxes will be the destruction of farming are talking nonsense. The land will always be farmed, whoever owns it, and nothing is more likely to fossilise the present structure and prevent the arrival of new blood, than prosperity and relief from capital taxes.

I have vented these sentiments before and have been roundly attacked for them as being a traitor to the side. But, quite honestly, I don't think any one family has a permanent right to anything for ever. Many years ago I was in hospital thinking I was dying and very nearly did. I was still more worried because I had at the time done nothing to alleviate the pressure of death duties which, at that time and at those land values, should not have been too difficult for an energetic family to cope with.

But at the same time I suddenly realised that whoever had to do the worrying, if the worst happened, it would not be me. From that moment I began to improve, and was extremely pleased that I had not given a great chunk of my property away. Although I have since got rid of a great deal, I have still enough left to do as I like without having to consult family or trustees. My wife too is covered by Mr Healey's concession that there need be no capital taxation between husband and wife. Many accountants are dismayed by this arrangement because it puts off some of their more complicated schemes of avoidance and thus reduces the fees they might otherwise earn. But I think it has been one of the best pieces of legislation that was ever enacted.

Of course the worst thing that has happened to farming over the last few years, has been the inflation in land values. They are, in many cases, self-inflating because investors – be they individual farmers or institutions – follow fashion. Because farmland has shown the best increase in value of any investment over the last 30 years or so, it follows, according to the pundits, that this must go on. Any farmer in search of expansion who owns some land uses it as collateral to buy some more, reasoning that the increased earnings will pay the interest charges. Then the progression goes on with increased asset value as security for yet more borrowing. It is as good a way of building a house of cards as any I know.

Not long ago, in 1977, some 100 acres came up for sale next door to me. It was a block of land which I had thought I had actually purchased for £30 an acre 30 years before, but at the last moment the vendor withdrew it. It was not good land, about 800 feet above sea-level and I did not like it much. But it was put to me that it should be bought. I rang the agent and he laughed to scorn my offer of £500 an acre which, at 14 per cent – the Mortgage Corporation rate – would have cost me £7,000 a year in perpetuity, and added £7 an acre to my total rent bill. At present this averages £10 per acre. Applied to the land itself, it would be a rent of £70 an acre. It eventually sold for just about double my offer, which would have cost me an extra £14 an acre over the whole of my acreage or, if applied to the acreage purchased, a net rent of £140 an acre a year. This is more than the value of the average crop of barley. I could, no doubt, have coped with this but it didn't seem sensible at the time and still doesn't.

It is said that English land is cheaper than that on the Continent, especially in the Community. This is not quite true. French land can be bought at very much less money, but this is because land purchase in France is very much restricted. There is a series of measures which make sure that it is reserved for actual French farmers who have no other land. German land is very expensive because, as one German told me, only a bloody fool would sell any. So the amounts coming on the market are limited. German farmers regard land as an investment akin to gold coins, with the added advantage that one can eat its produce.

In terms of what can be earned from them. British rents are still cheap in comparison with those in some overseas countries. In New Zealand, a young farmer with a herd of cows can rent a farm for half the milk cheque and a proportion of the calves and pigs reared. This is because the demand for farms, even at the low returns in New Zealand, induces young men and their wives to make any sacrifice to get into farming. In Australian share arable farming the owner can command up to 40 per cent of the crop, in some cases without putting anything in except a little superphosphate.

I believe that in the future, share farming deals of this

sort will be one way in which young men can get into farming. Although an old man myself, I have taken over a neighbour's farm which I can work with my own tackle for a limited period, because he wished to give up the land but not the house and the freehold. His share is his pension and is taxed as a trading profit because he does contribute the fertiliser and some other inputs and carries some of the risks. It is an excellent system in that he has to suffer or enjoy the fluctuations both of yield and price in a way in which a landlord under a straight rental agreement does not.

A farmer, whom I know, keeps a large herd of pigs on another farmer's farm on the outdoor system. This is a good arrangement as the pig farmer does not have to have very elaborate and expensive buildings and his capital is invested in productive assets much as François Bouchardy suggested. I would be the first to agree that my old friend's advice about investing in land itself is rather beyond sensible reach these days. I was fortunate in being able to follow it when land prices were more reasonable

But there would still be scope for reducing investment in machinery by hiring it and even better, to hire contractors to save the cost of employing labour and machinery at all. Really the problem is to make use of the money available so that it will earn the greatest return while maintaining itself as capital and an increasing asset. Once spent it is gone for ever, so it should always be spent with caution.

At one time I was very seriously strapped for money and it was suggested that I should sell some of the land I had bought and lease it back over a period of years. This was when £100 an acre was a good price. It looked tempting at first, but the only purpose in effecting a lease-back would be to earn more money by a higher turnover and thus make myself liable for more taxation.

More seriously though, I would have lost my equity – the basic worth of the land on which I could always borrow more money if I really had to. I know many farmers who have expanded their businesses enormously by leasing back their land and getting, as they believe, more in the way of income from it. But today most of them are on balance worse off than they would have been had they not sold out. Circumstances alter cases and what is right for one generation will seldom fit the problems of the next. But the essentials of François' advice – to avoid investing in assets which are fundamentally wasting – must be right.

22. Feeding the World

In the late 1960s, the B.B.C. was running a series called 'One Pair of Eyes', in which an individual recorded, with the help of film, his thoughts, fancies, pastimes, philosophy or his obsessions. I had an obsession at the time, with which I used to bore the pants off anyone who would listen, and will do so again at the drop of a hat. This was simply that those who agonised about the population explosion and consequent world starvation were wrong; that the world has ample resources to feed a population several times that of the present day, and the fact that it fails in certain countries is entirely due to the mismanagement of the countries concerned. These sentiments which I had published several times, were objected to quite violently in places like the F.A.O. in Rome, where the threat of mass starvation has kept hundreds, if not thousands of international Civil Servants in tax-free jobs for many years.

I wrote to the producer of the series, Anthony Lotbiniere, whom I never met incidentally, with a synopsis of this theme and he obliged by putting the wheels in motion. Room was found in the budget; a director – Richard Marquand – was appointed; and I was on the way. My basic thesis was that, given sufficient resources of cash and knowledge, there was no problem of food production which could not be solved within the boundaries of almost any country. To illustrate this, I suggested looking at what had happened to the derelict Hampshire farms after 30 years of good farming; to the American Middle West, once a dust bowl; and to countries in South

America where production was falling steadily because of the lack of fertiliser.

I had been very much influenced in my ideas by the work of Norman Borlaug of the Rockefeller Foundation, which work was centred in Mexico. Borlaug is one of a rare breed, a practical scientist. He had been sent to Mexico after the war to increase Mexico's grain output by breeding new varieties and developing new techniques. His problem was essentially a simple one. Nitrogen was the key to the increased yields, but existing strains of wheat were too weak in the straw to stand any increased application. In addition, the wheats had to be able to stand subtropical conditions and have reasonable bread or pasta making qualities.

Mexico City near one of his stations is at about 7,500 feet above sea level and in a relatively temperate climate. Cuidad Obregon in Sonora, where he had another research station, is subtropical. So he was able to accelerate his breeding problems by having two harvests in one year. Within about 20 years he had developed by selection the dwarf wheats which formed the basis of the Green Revolution. The Foundation has also developed a high-yielding maize and the Philippines rice which practically doubled yields.

It has also demonstrated that the new varieties could be used without necessarily demanding a whole host of scientific and mechanical assistance. This was the result of the Puebla project, applied to a number of villages on the slopes of Popacatepetl not far from Mexico City. This is an area of smallholdings, extremely poor and overpopulated. Crop yields were minimal, and the Rockefeller team were instructed to develop systems of husbandry using the peasants' own resources: oxen, mules, donkeys, even human muscles. What they taught were simple husbandry techniques, water conservation, and the right dates for sowing the main crop which was maize. Yields then doubled and trebled to approximate those of well-farmed land.

Borlaug was not as foolish as to suggest that his were the answers to all the world's problems. But he maintained, and I agree with him, that if the means of feeding populations are on hand, it is then up to the politicians to solve the other problems of sharing and general organisation. The Green Revolution has been viciously attacked by many interests, both left and right wing, on the grounds that it destroyed the independence of the peasantry, and made them dependent on the products of industry, particularly nitrogen. It seems that these people wish to deny the peasant classes, for want of a better term, the benefits of freedom from starvation.

The successes of the Green Revolution have been enormous. Starvation is now either a matter of climatic accident or mis-management. The first can be cured by aid from the world's growing surpluses, the second by organisation within the country concerned. There is evidence of widespread malnutrition in India for instance where wheat production has doubled over the past 20 years, thanks to the Green Revolution. There is now in fact a surplus. In China – a land of much greater population pressures on the available arable land – there is no evidence of malnutrition. Nevertheless, the Chinese quickly took advantage of the Green Revolution strains. This may not be a reflection of the difference between Communism and democracy, but of the relative farming and political skills of the two nations.

What made the Pueblo project so important to my mind was that it attempted, with some success, to combine social justice with increasing agricultural efficiency. The Pueblo peasants were the beneficiaries of *Ejido*. This was a system of land reform initiated after the last of the Mexico's great revolutions in 1917. The great estates were broken up ruthlessly and the land given to the peasants in holdings of no more than 10 acres each. It is obviously better to farm in big units to gain the efficiencies of scale, but the Rockefeller Foundation decided this was impossible, and so they deliberately set out to improve the smallholding system.

I have studied land reform in many countries over the last 20 years and, while the need for a redistribution of wealth is obvious, simply to hand out a few acres apiece to each peasant family is a certain route to bad farming and reduced production. The Chinese have got over this by forming the communes and by having a system of farming which, while efficient in production per acre, makes an inordinate use of manpower. Most of the countries engaging in land reform have tried to avoid the worst of the inefficiencies of fragmentation by attempting some form of co-operative organisation. Having right-wing or even socialist governments they did not wish to go the whole Communist hog.

In Colombia I visited some land reformed holdings where I attended a meeting of the new owners. Several officials from the office of the Reforma Agraria were trying to persuade a group of peasants of the benefits of putting their land in common. They would grow better crops and be able to sell them better. At the end of the meeting a pretty violent-looking individual got up and said to loud applause that the Colombian countryman had not shed his blood in countless revolutions since the ending of Spanish rule only to give away the hectare of land which his sacrifice had earned. He was loudly applauded, and I felt quite sorry for the resident administrator of this particular estate who would have to try and persuade these savage-looking people to merge their identities for the common good.

Colombia is not so badly off for food. In the tropical lowlands I met men who raised quite large families on less than a hectare, and in the hot climate there is no demand for warm clothing and housing. The problem here is to create enough trade and industry to make sure that the benefits of industrialisation reach the country. Colombia is fortunate in its coffee; many thousands of peasants get a minimal cash flow from the produce of a handful of bushes. Marijuana is an even more profitable crop.

Peru is a very different story. West of the Andes it is a desert with some irrigation; in the Amazonian jungle, development is not far advanced. Over the last 10 years the military government has instituted some quite effective land reform. The farms are largely co-operatives and the officials I met were most effective. Stock and crops were good and the farmers seemed happy. The set-ups were very disciplined, and I was not surprised to learn that they only approved candidates for smallholdings who had never had an independent enterprise, however small, before. Such people, I was told, were apt to go to market and get drunk. If they had been too poor to get into either of these bad habits, they made much better settlers. Incidentally the administration also had a great say in the family diet. Every housewife had a box of guinea pigs in the kitchen. They eat the scraps, and the family eats them to provide protein. The Government maintains a special research station for improving the guinea pig strain with weight gain recording and feeding trials.

Peru has an appalling poverty problem and, while land reform was making quite a contribution to food resources, it was doing very little to overcome the basic insecurity of the masses. I have often thought that the slums of Lima are among the most desolate in the world; particularly as, in the desert environment, no trees or other greenery grow as they do in other tropical slums.

I was in Chile just before President Allende was killed and visited quite a few of the legally expropriated estates. As a government guest I was, presumably, shown the very best there were – they were a shambles! There was a complete lack of practical knowledge among the leaders of the settlements who had obviously been appointed as politicians and not farmers. The only settlements that still worked were those where the previous owner had remained and advised his ex-employees what to do.

One of the land reform officials was very upset because he found that, when they had created a settlement on co-operative lines, the members – who were usually the workers already employed there – refused to allow any of the local landless peasants to join them in case it diluted their profits. They even refused them employment when they were busy. It seems as though the spirit of capitalism soon comes to those who have something. Chile is a tragic case. Most of the better land is in the temperate zone and living there does require good clothing and heat. The land, though good, has not the productivity of the tropics either.

As a direct contrast at this point, it is worth mentioning an effective means of land reform which operated in the district of Argentina where I used to work. The Code Napoleon which fragments the land of Europe by insisting that a father's property is divided equally between his children on death, operates in much of Latin America too. It is one of the few laws to be kept seriously because every heir has an interest that it should. When I worked in Argentina, the company farmed 100,000 acres for the benefit of three individuals. Now, 50 years later, the 100,000 acres has more than 60 owners. In the old days, the output from the ranches was not more than a few thousand steers which were shipped to Buenos Aires for slaughter and export. But this meant that the little towns that serviced the estancias, and others like them, were no more than railway stations and a few stores. The principal town, Ameghino, only had paved roads round its central square.

Beef cattle, under any form of farming, are the least productive converters of resources into money. Once the farms were made smaller, the owners had to turn to cropping and dairying. This meant silos, milk factories and other ancillary trades. The population of the town grew, and it is now quite a sizable and prosperous place. The same goes for much of the better land in Argentina.

Argentina is fortunate that an accident of history gave it a vast area of good and fertile land in a fairly reliable rainfall belt. Although the inhabitants are not very good at governing themselves, the Argentinians have been able to prevent the rich from robbing them to the point of starvation. This is one of the few Latin American countries where social injustice is not blatantly evident. But Argentina is far from being able to fulfil its production potential. Fifty years ago its grain and beef dominated world markets. Today these exports are well below those earlier levels. Part of this is due to an increased population, but the main reason is that no fertilisers have ever been used.

This is because of the insecurity of all world food supplies. Farmers have not been able to earn enough to make the use of fertilisers and improved farming systems worthwhile. Or if they have, as they did at one time in Argentina, the then Government took it from them in taxation. This isn't just the problem of Argentina. There are countries whose only possible asset is the export of primary products. Every time a rich industrialised country becomes self-sufficient in food, as for instance the Common Market is attempting to do, it removes the opportunity for export from some much poorer country elsewhere.

This is far from a fanciful notion. The European Community is doing this in several different ways. Its subsidised exports of sugar are not only taking the markets of the rich Australians, but of the Jamaicans and Mauritians. It is attempting to limit, or prevent the importation of Thai manioc in order to protect its grain growers. Italian citrus growers are protected against imports from Israel and Spain. The Community is not the only offender by any means. The United States, while crying foul at Community protection, is highly protectionist itself.

I came across another aspect of this in Fiji some years ago. I was taken to see a banana plantation which had been started by the Commonwealth Development Corporation, partly as a social relief measure. It was a good job: each settler had about ten acres, five acres of which were planted for him. It was all very impressive. The next day someone else took me out to see a long-established banana farm, where the fruit had fallen and been left to rot on the ground. There was just no export market.

It is a great pity that social justice has in many countries been equated with land reform, with all the harm that this does to common-sense economics. In fact, any attempt to solve this problem by merely handing a family a few acres of land, will probably decrease the land's productivity in the short, and even long, term. There are many underfarmed areas in the world, of course, and these should be sensibly tackled.

There is far more at risk from the political threat of increasing populations than of their starving to death. Man does not live by bread alone, but how do you give sensible occupation to the millions of young unemployed in some of these countries? In Mexico for instance, between 30 and 40 per cent of the population is workless and, with the country's current population growth, this will mean 30,000,000 or 40,000,000 before too long. In the smaller Jamaica, about half the men are out of work with no hope of emigration.

Even in the United States and Europe, unemployment is serious and once the rulers of China lift their millions of workers from the grindstone of subsistence farming, they too will find – if they haven't found already – that their educated young people will be questioning the basis of their existence. With modern communications, even the language barriers would be swiftly overcome in some spontaneous reaction to what many would feel to be the injustices of their existence.

This I feel is the basis of Borlaug's philosophy. In my interviews with him, we went through the physical problems of creating more food, and of overcoming the lack of knowledge of good farming practice, which is mainly exercising common sense. What he said he was doing with the Green Revolution, was providing the world with a breathing space, during which the problems of feeding the increasing populations could be of lesser urgency, while Governments and nations were tackling the much more fundamental problem of existing in a modern society.

So far no one seems to have given the least thought to this dilemma, so obsessed are most people by the Malthusian thesis. In case you have forgotten, Malthus prophesied 200 years ago that, within a measurable space of time, the world would run out of food resources. There is no chance of that happening and Malthus has been proved wrong many times over. Ungovernable discontent is far more likely to finish us off.

23. Family and Friends

If you have read so far, you might well be wondering why I have kept working into my seventieth year and fully intend to go on doing so as long as I possibly can. Broadcasting and journalism will undoubtedly fade, although I have enough work sold to keep me going into my seventy-second year. Then I still have my farm, at the moment about 1,000 acres. This is fairly intensively run; I have complicated my life by having not only a large flock of sheep, but also the best part of 150 sows.

Uncle Jack used to say that the best motto for a farmer was: 'Live as if you were going to die tomorrow, but farm as if you were going to live for ever.' Apply that to journalism as well and you have a recipe for never letting the young into anything. If, instead of farming, I had gone into paid employment and reached the highest levels in, say, B.P. or a bank; I should have had to retire at 60. Being self-employed does not provide for an indexed pension, which is as good an excuse as any for not retiring.

I have also been very lucky indeed. Any farmer with four sons these days is someone to be pitied, but I managed to acquire enough land so that I could start them all off and go on farming myself. The fact that I have so much still – 1,000 acres is quite a spread – is because Tom, my one from last, has chosen another trade and lets me the farm which is his share.

My life would have been much eased if there was a way of controlling the sex of one's offspring on conception. My ideal family would have been – if we had had five, which was too many in any case – four daughters, then, a son, as an afterthought. He would just about be ready today aged about 25 to help me out on the 3,000 acres I was farming, preparatory to taking over the lot in 10 or 20 years' time.

My daughter Susan encouraged this view of the benefits of a female family. She never gave me the least anxiety, playing quietly when young with her dolls or whatever girls play with. She worked her way to Oxford and a good job; then married a splendid husband. She did show signs of the character which could keep comprehensive pupils under control. While on her teaching practical, she used to come home from her North London school with a bag full of toys which she had confiscated and gave them to her younger brothers. With her, firmness means just that.

Once when Ronie was ill, Susan was looking after us all. One evening she drew me aside and laid down an ultimatum.

'Father,' she said, 'if anything happens to Mother, I will give you six months in which to find another woman. I am not going to spend the rest of my life looking after this lot.'

How right she was! Farmers' families are full of the sacrifices made either by sons or daughters who gave up the best years of their lives to look after their families.

I remember a well-known Wiltshire farmer dying at about 70. His mother, then aged 92, departed this life a fortnight later: the object of her living, whom she had looked after from his birth, was no longer there. I used to quote this to Ronie as an example of duty above all else. I accept that if the need had arisen she would have coped; but after the boys were about 25, she said she would start issuing ultimatums and parading prospective wives on an 'either/or' principle.

Undoubtedly my business would have been much more efficient if it had been kept together and run as a large integrated company. This could have been done and the enterprise could, by the skill of the accountants, probably have been kept intact. Although all the boys have worked for me – Rowan the eldest has in partnership for a while – in the end they went on their own. I do know many very successful family businesses in farming and elsewhere, but those involved must have the patience of saints. Many have told me that, while the men of such an arrangement get on well together, the whole relationship can be ruined by their wives. This is a typical male attitude.

In any case the finest thing anyone can give his son is the opportunity to be his own boss. All mine were on their own by the time they were 24 or 25 and, once they were on their own, I never went on their farms except as a visitor. They would be bound to do things which I would not do, and think of the frustrations saved on both sides by this separation.

Of course the fact that I am fanatically for self-employment may have coloured their upbringing. Also my misery at Berkhamsted made me think that all schools today are just as bad. I am assured that they are not, but simply don't believe it. My belief is that the disciplines of a good home are far superior to what can be acquired in any boarding schools. As an executor, uncle and general observer, I have seen that, in the present fashion for dropping out, public school children seem to fall faster and farther than those of day schools.

I claim no credit for this at all. Ronie ran my home around me with firmness and commonsense. I say around me, because I confess I was always too busy either with farming or other ploys to devote myself as I should to the well-being of my family. I was always a good provider, but I had also the good fortune to be able to detach myself from my surroundings while I attended to the more important things in life. This trait has been invaluable in open plan newspaper offices, and television and radio studios. This is one of the most useful attributes to possess and one of the most selfish. I have ruthlessly cut myself off from conventional family life, although I love them all dearly, to pursue singlemindedly farming, writing, journalism and other things.

Writing is probably the most obsessive of these. I write two or three articles a week for various outlets. Generally speaking they come 'off the top of my head'. Unlike journalism, which really is reporting, writing is for want of a better word, an art form – at least to some extent. It doesn't take me much more than a couple of hours to write 1,000 to 1,200 words; if necessary I can do it in half the time. But thinking about the subject does take longer. Ronie knows the symptoms well. I don't listen to the conversation, forget engagements, and generally behave as if my mind is miles away, which indeed it is. Some would call it the search for inspiration. I wouldn't call it that at all. Inspiration is for great artists, but I always have a whole kaleidoscope of ideas seething in my mind. Every now and then they suddenly crystalise into a theme; if the theme is apposite to the moment, there is my next article.

You can understand poor Ronie's problems – almost from when we were married, if not before, my mind has always been in such a state. I suppose most people's are. The lucky ones can apply them to pure works of the imagination and become great artists of every kind.

A very shrewd friend of mine once said that my obsession with writing and broadcasting was a form of escape. In a way I suppose he was right; it is not difficult to see that I was escaping from my family obligations as conventionally understood. For this I must blame my upbringing. I really resented the degree to which my will was subordinated to that of the family. I used to think of this and so rationalise my own attitude to my own family.

I sometimes think it is a great pity that my mind's kaleidoscope is filled not with ideas of great moment or with plots for novels and plays, but with the more mundane matters of farming and the Common Market and human behaviour. I read a great deal. But by the time I have finished *The Times* and the *Financial Times* before breakfast, I am already sunk in contemporary affairs. I follow these papers up with numerous magazines and newsletters, all of which deal with essentially the same themes. I have a good memory for ideas and if figures are going to be needed, they are kept on file.

The ability to read is the greatest gift and, as children, we could all read at an early age. But we were made to read good books, not improving books but good books. This compulsory reading was carried further at school with the classics and Shakespeare and turned me off such reading once and for all. I read trash for years until, all of a sudden, I didn't want to any more. I still read detective stories but mainly biography and travel. The only field sport I really enjoy is fishing. I used to shoot a good deal, and was at one time a fair shot. Then I went right off it. I don't mind walking round with a dog and a friend and shooting at a few birds, but formal shoots leave me cold. I could never be bothered to rear a lot of pheasants and, although reared birds are not as easy to shoot as the critics say, there is something artificial about it. I must admit, though, that when someone else's reared birds come over the boundary, I don't feel the same way.

I used to do a little trout fishing in Ireland with my father, or when we took the children to Wales. Then I began with the dry-fly in a desultory manner; but I was always prepared to leave fishing for work at any time. Then came a moment of truth. Maurice Jones, who managed the Leckford Estate, had given me the odd day's fishing on the Test and some 20 years ago in hospital I had a letter from him wishing me well. I suddenly thought that I might die and never know the delights of real dry-fly fishing. So I wrote back and asked for the first reversion of a rod on the Leekford water. This meant that I could give even less time to my family responsibilities but, whatever they suffered, my writing did not.

Dry-fly fishing is made to sound difficult, but it isn't. The idea is that you wait until you see a trout rising to the hatching fly which is drifting down the river; then you drop your artificial fly in front of it. The skill is in the casting: once that is learnt, the rest is comparatively easy. It's not difficult to learn to cast either. After one year, I won the championship at the Game Fair. Modesty is not – as you must have found by now – one of my strongest points, but I did become quite a good dry-fly man. Then two things happened: I began to feel sorry for the fish I caught, and I got bored with it. At least I didn't so much get bored, but met another way of fishing. I had done a little salmon fishing with only occasional success when I suddenly became obsessed with the determination to catch a few more. It is just as ridiculous an obsession as gambling or drug-taking. There is no skill at all in it: the veriest tyro has just as much chance of killing a fish as the absolute expert. No one knows why the salmon, which never eats once in fresh water, should take a fly or any other bait. On the occasions when the fish are taking, they will go for anything for an hour or so; and then they will sulk for the best part of a week.

I have stood in water up to my waist for a week casting every 30 seconds, watching the fly come round, lifting the line and casting again. I never get tired of it and I go whole years without catching anything. But one of these days before I die, I hope to have a real good day – half a dozen fish – then I would go content.

24. The Essential Factor

I would never have made progress in farming without the help of the men I have employed through the years. I am not one of those, however, who nostalgically extol the inherent skills of the countryside. There certainly were craftsmen but, when I started, for every craftsman there were at least half a dozen farm-labourers. These men could hack down hedges but not really lay them, stook and pitch sheaves, spread dung, and help in building hay ricks. They did work which any intelligent men – and a lot of them were intelligent – could learn in a very short time.

The aristocrats of the farm were the head carter, the cowman and the shepherd. But the shepherds who used to rule the roost on the arable farms of Hampshire and Wiltshire were already a dying breed. Sheep no longer paid and their dung could be more cheaply replaced by bag fertiliser. Cowmen were an unpopular necessity, only dairying really paid, and they were beginning to rule the roost on many farms. Carters too were on their way out. Some of them took to tractor-driving and became very good at it. The best of them had an instinctive skill in working land and were able to adapt what they had learnt behind the plough while in close contact with the soil, to the tractor-seat. In general though tractors were taken on by the younger men.

I had two horses when I started at Knoyle; Duke which I bought at the outgoer's sale, and Flower which I got from Jack Stratton. I also had a tractor and bought a brand-new Cockshutt plough as the old tractor plough purchased at the sale seemed to have a twisted beam. It always had had, I was told. I had taken on the under carter, Jack Bowing and I used to hear him tramp up to the stables at 5.30 a.m. as I was getting ready to go milking. He used to feed, groom, and harness his pair, and be ready to start work at 6.45 a.m. He took to driving the tractor, a Fordson, like a duck to water but, like me, had never ploughed with one. Ploughing is not really difficult once the field has been opened out, so after making a fair mess of the first field, he used to harness up his team and opened the furrows with it. He used it to close them too, the other ticklish part. This went on until the agents sent out an expert who showed us how to do it with the tractor.

Jack worked for me for many years and graduated to using caterpillar tractors, five furrow ploughs and to driving combines. He became a first-class operator and highly skilled in all sorts of work. Like many intelligent farm-workers, he was difficult at times and always carried a chip. He felt, as many of them did, that they were stuck with farm-work, and were missing the better things of life. Stuck they most certainly were. There were two reasons for this: the tied cottage and rural education of the time. The tied cottage certainly kept them on the land, because it made it effectively impossible for them to leave farming, at least once they were married. But I have always believed that, with a few exceptions, rural education was much more responsible.

There were good and bad village schools but, in general, the standard of education was bad and they were under the eyes of the parson. In turn he was under the influence of the wealthiest of his parishoners: the farmers and landowners. People often deplore the decline in the influence of the Church of England. But whatever influence it once had with the masses was lost by its attachment to the propertied classes. Nowhere was this more evident than in the country. Until the end of the last war it was almost impossible for a child from a village school to move on to secondary education. Children were born to be farm-labourers or other rural workers and few school-teachers pushed their children to the grammar schools which could have taken them. Since all secondary education has been concentrated in the towns and country children finish their time there, the standard of young people coming to work for me has improved immeasurably. They have broader minds and have seen something else of the world. Then they have come back to farming willingly.

The present farm-worker is now a skilled man, doing himself the work of five or six of his forbears in every skill. Is he well enough paid for this? And if not why not? I don't believe he is for a variety of reasons. His position is weak. Unlike workers in mass industry, his relations are much closer to his employer and not with his shop-steward. Most feel responsible for the stock and crops they look after. Many are on Christian-name terms with the farmers and are often the only man on the farm. It is difficult to become militant in these circumstances, particularly when living in a tied house.

Until the Agriculture (Miscellaneous Provisions) Act of 1976, falling out with an employer meant the loss of a home. This new measure laid an onus on the Local Authority to provide, in certain circumstances, accommodation where the outgoing worker's house was needed for another essential worker. Although the N.F.U. attacked the measure, and the Farm Workers' Union claimed it did not go far enough, it does seem to be working. The tiny number of actual evictions has fallen even further.

But his position is also weak because farmers are still getting rid of labour. It must be remembered that, partly, because of farm size, British farmers employ more workers per 1,000 acres than any other temperate farming in the world. This has mainly been due to the fact that land has been historically cheap as have the costs of other inputs. Land, tractors, fertilisers and chemicals are essential farm inputs which cannot be dispensed with. Farm-labour has been leaving the land steadily as farmers feel the cost squeeze and seek to alleviate it in the only way left to them.

It has been a long process. In 1946 I was employing, according to my wages book, 22 men altogether on 1,650 acres. Part of this was due to war-time farming and the need to grow potatoes, sugar-beet and flax. Most of the harvesting of these crops was done by gangs. But six men used to be employed riddling out potatoes, fencing, etc. There was a maintenance man and his son for work on buildings, a mechanic, and half a dozen tractor-drivers. I had a herd of cows and a flying flock of sheep, but most of the work was to do with the arable.

Today, in 1979, I have a permanent staff of seven on 1,000 acres. They are: a working foreman, employed because I have other interests; a maintenance man; two tractor-drivers; a shepherd; and two pigmen. 650 acres of grain produce 1,200 tonnes of corn; the ewes produce 2,000 lambs and the pigs theoretically the same number of porkers. I employ casual labour for a total of about three months through the year at lambing and harvest. If I was being squeezed, I could dispense with the foreman and the maintenance man. Many farmers in my position are employing far fewer than I am able to afford, thanks to buying my land at an historically low cost. There is no problem in finding workers. An advertisement, even under a box number as I have used over the last three or four years, brings up to 30 replies, most of them good. One in the local paper brought a boy to start with the pigs by 8 a.m. on the morning of publication.

There are obviously men who like farm-work, as evidence a number who apply from other jobs and who own their own houses. I am always telling school-boys to leave the industry alone. The obvious disadvantages of the tied cottage are outweighed by some of the tax-free perks they get. I don't think farmers should claim too much credit for these. My cottages are worth between £15 and £20 a week on the open market inclusive of rates. The reason farmers forgo these advantages is because they need the men to live near their work.

The Farm Workers' Union makes a big thing of the workers being always at beck and call, liable to be called out at any time, especially if they are in a tied cottage. But the simple answer to this is that taking up farm-work is voluntary. A boy today has always the chance of going to work in a town if he has been educated there. In any case, there are no opportunities in the country outside farming and precious few in farming.

My good fortune in farming has been largely due to Stan Nokes. He worked for me for nearly 40 years, most of the time as foreman. He had left school at the age of 11 at the outset of World War I which caused quite a rush from the land. There were no reserved occupations in those days. It must have been a vintage year because another foreman I know was of the same generation. His first work was as a carter's boy and he graduated up to being a cowman on a bail.

Start always had an independent attitude and fell out with the farmer's son where he was employed. The farmer, knowing I wanted a man, recommended him for a period, saying that he would like him back as soon as he could get some sense into his son. Good cowmen, he implied, were much harder to find than sons. Stan was a very good cowman indeed, but always told me that when arable farming was able to pay better wages he would like to go back to it. He issued an ultimatum at the outbreak of war: 'Put me on the arable or I leave.' Obviously had there been the Eleven-Plus in his school days, farming would never have seen him. Nor would he have stayed long with me, had I not put up with his independent spirit and contradictions for the sake of his skills and loyalty to me and my family. His son, my present foreman, would also have gone a long way, had the educational system been different when he was young.

I don't think a good man can lack independence of spirit; and while there was a certain amount of low cunning among the countrymen years ago, there was not much overt independence. I had a theory that the independents had all gone to the towns over the generations, where there were other opportunities.

Of my present staff, three have been with me more than 20 years at least. The others have mostly replaced men who left to retire or by mutual agreement. One very good tractor-driver wished to better himself and became a foreman. I told him there were no prospects with me as I already had one, and had my sons coming on. I recommended him for his first and second foreman's jobs, and he is now farm-manager in all but title on a very big farm in the public sector. Harold, for that is his name, would undoubtedly have become a farm-manager had he had a degree or diploma. But, if he had stayed on at school and then gone to college, he would have missed the early training which made him an exceptionally good practical worker. This is something no formal education can provide and given the choice I would always engage the practical man.

Maurice who was with me for 12 years or so was a man with a superlative eye. He could drive a tractor straighter than anyone I knew. He never needed marks and he never overlapped when spraying or manure distributing. But he had a wife, and she was determined to get on in the world. Her chance came with the offer of a post office in the next village. For this my neighbour could offer a house and so Maurice gave notice and left. When he told me about it he said he was sorry, 'But Mr Cherrington you know what women are.'

I have only had to get rid of two difficult men, both of whom I had inherited when taking over another farm. This was not so much a matter of incompetence, but of mutual incompatability on both sides. The first essential in the relationship between employer and employee should be that they are prepared to work together and to make allowances. This may sound a counsel of perfection, and I suppose is impossible in large-scale industry – but there is no alternative to it in farming. Don't think there is anything idyllic in my relationship with the workers, or in any other farmers', some of them have very bad names as employers.

We only employ men to make a profit out of them and, however well we pay and treat them, we spend no more than is necessary to get them. I couldn't say how much more we could pay because that has not yet been put to the test. For instance in industry I understand that, when an employer puts in a new machine which will double the workers' output, the worker or his union demands a share which sometimes will completely nullify the benefits of the investment. This is not so in farming. However, we shall have to pay them more if they are to keep up with the urban Joneses. A decent second-hand car used to cost £200–£300. Such a car now costs over £1,000 and is essential to take the wife shopping or out to work, which many of them do.

I used to advance the money needed to buy a car at the lower rate: but will they be able to afford such a car in the future without substantial wage increases? The problem is not only the wage-scale, but the tax and insurance which are deducted from even quite a low wage-packet. This is what really makes them sour and drives many to work elsewhere in their spare time.

It is not sensible in a time of inflation to estimate future wages costs, but a good cowman can already command £100 a week and there is no doubt that the present union claim of £80 a week minimum will look very moderate in a few years time. Perhaps the magic figure of £100 will come in my life-time. Farmers throw their arms up in horror if you say this to them, but it will became a fact of life and will have enormous repercussions on the future pattern of farming, as will the increasing price of land.

Anthony Rosen, who built up Fountain Farming from an idea to some 27,000 acres in a dozen years, always maintained that his would be the pattern of the future. Only by becoming a member of a management team could a young man achieve his ambition of farming a worthwhile acreage. While giving full credit to Rosen's ability, and to that of his imitators of various kinds, I think he and they are wrong. His argument is that land prices have risen so fast that only institutions will have the capital either to own or farm any worthwhile acreages. Taxation has wiped out the individual small farmer down to quite a minor acreage. They do not think that any young man of spirit will be prepared to condemn himself to a life of drudgery on a small acreage just for the sake of remaining independent. But it all depends on what you mean by a young man of spirit.

It should be realised that, in sizes of farm-businesses in the temperate areas, Britain is the odd man out. In the Common Market, in the States, in Canada, Australia and New Zealand the family farm is paramount. Some of the farms are tiny like the part-time farms of Germany; others, like those in the U.S.A., are very large. The interesting thing is far from the farms becoming larger and employing labour on the British model, they are certainly becoming larger but still remaining of family size.

Labour is very dear in all the countries mentioned. A tractor-driver in Germany earns at least £80 a week, a cowman nearly double that. The result is that fewer and fewer are employed because the margins in farming, even with Germany's inflated returns, are not enough to carry them. Already farm size is being limited by wage costs as well as land and other costs. There are very definite grumbles from the Rosen type of organisation that returns arc not sufficient to cover all costs and show a satisfactory profit, particularly under management.

But it is not reasonable to expect farm prices to rise much further in the short, medium or even longer term. If the C.A.P. blows up, prices are more likely to fall than to rise. Each country will protect its farmers as best it thinks. I doubt very much if the big unit has got a real future in economic terms at least.

Take poultry. As I understand it, in much of the E.E.C. today both broilers and layers are entirely in the hands of small, sometimes vertically integrated, units in every one of the member countries but Britain. This is not because the major operators like having to deal with small individuals, but because they are more effective and, in a bad time, cheaper than employed labour. I am sure that when Sir John Eastwood sold his empire to Imperial in 1978, this consideration was very much in his mind.

When the margins in dairying begin to narrow still more, it will almost certainly revert to the family farm or the share milker. I find it difficult to believe that pig production will be possible in the large labour-employing units, unless prices can be stabilised at a higher level than they have been able to be in the past. Only in arable farming could there be some possibilities of scale – simply because of the enormous sums needed to buy and equip even a moderate-sized farm. Even so, I believe smaller individual operators will be more likely to work themselves in. So certain am I of this development, which has been apparent elsewhere, that I would bet on it happening eventually.

This change will mean the end of the line for such as Stan, Maurice and the rest. They will be replaced by men seized by the desire to farm on their own account as far as is possible. They will make whatever sacrifices they feel they can afford to achieve their object. Some will probably succeed and their qualities, besides technical ability, will be that of being able to live, if necessary, on the smell of an oily rag.

OTHER FARMING PRESS PUBLICATIONS

The Hired Lad   IAN THOMSON
A young man's first work on a Scottish farm when horses were yielding to tractor and bothy life was rough and ready.

The Spacious Days   MICHAEL TWIST
Growing up on a Buckinghamshire estate in the 1930s. Anecdotes about the farm staff, agricultural work, gamekeeping and the countryside.

One Dog, His Man & His Trials   MARJORIE QUARTON
Sheepdog Shep's tales of his life in Ireland with its rogues, adventures and humorous encounters, canine and human.

They All Ran After the Farmer's Wife   VERONICA FRATER
The true story of a farmer's wife coping with seven young children, bed and breakfast and holiday cottage lets.

HENRY BREWIS
The six books by Northumbrian farming humorist Henry Brewis include three cartoon volumes, stories, poems, a diary and *Country Dance*, the story of a hill farm.

JOHN TERRY
John Terry's three school farm books humorously chart the growth of a rural studies department from wasteland to prize-winning sheep unit.

JAMES ROBERTSON
Two 'Any Fool' titles show the pitfalls awaiting the newcomer to pigs and dairy farming.

For more information or for a free illustrated list giving full details of our wide range of agricultural and veterinary books, please contact:

**Farming Press Books & Videos, Wharfedale Road
Ipswich IP1 4LG, United Kingdom
Telephone (0473) 241122   Fax (0473) 240501**

Farming Press Books is part of the Morgan-Grampian Farming Press Group which publishes *Arable Farming, Dairy Farmer, Farming News, Pig Farming, What's New in Farming*. For a specimen copy of any of these magazines, please contact the address above.